AFRICAN ENVIRONMENT
SPECIAL REPORT 1

AFRICAN ENVIRONMENT

Problems and perspectives

Editor Paul Richards
assisted by Nicola Harris

Published by the
International African Institute
in association with the
Environment Training Programme
UNEP-IDEP-SIDA

INTERNATIONAL AFRICAN INSTITUTE
210 High Holborn, London WC1V 7BW

ISBN 0 85302 046 9

The opinions expressed in the text are those of the contributors and not necessarily those of the IAI.

CONTENTS

*Originally written in French. The remainder were originally written in English.

ACKNOWLEDGEMENTS

The editor would like to thank the long and short term members of the International African Institute research staff for their help, criticism and advice: Fatima Bezzaz, Isabel Boothman, Abdul Mejid Hussein, Julius Oguntoyinbo, Clare Oxby, Abdul Karim Turay.

Redrawing Maps and Figures

Alick Newman, Department of Geography, University College, London.

Illustrations

We should like to thank the following for their kind permission to print photographs and figures: the National Oceanic and Atmospheric Administration, USA (Fig. 2.1); P.H. Kuenen, and Wiley and Co. Publishers, New York (Fig. 2.2); A. Rapp (Figs 2.3;4;5).

FIGURES AND PLATES

CONTRIBUTORS

LENNART ANNERSTEN is Administrative Director of the Mathematics/Physics Department of Uppsala University.

MACHIOUDI DISSOU is Lecturer in the Rural Economy Section of the Département des Etudes Agronomiques et Agrotechniques of the University of Dahomey where he has made a particular study of the palm-products industry.

JEAN DRESCH is Professor of Geography at the University of Paris VII and has a long-standing interest in the geomorphology and environment of arid and semi-arid areas of North and West Africa and the Middle East.

MAHDI AMIN EL-TOM is Head of the Geography Department at the University of Khartoum and has specialised in tropical climatology, especially in the Sudan.

JOHN FORD has long experience as adviser to various African governments on tsetse fly problems.

JOHN J. HIDORE is Lecturer in Geography at the University of Khartoum.

PHILIP LANGLEY is a member of the Environment Training team based in Dakar with a particular interest in the educational and perceptual aspects of development.

BENGT LUNDHOLM is Lecturer in Ecology at the National Board of Health and Welfare and Director of the Secretariat for International Ecology, Sweden.

PHIL O'KEEFE is a geographer with research experience in Kenya now working with the Disaster Research Unit, Bradford University.

LEKAN OYEBANDE is Lecturer in Hydrology and Water Resource Management in the Geography Department of the University of Lagos.

PAUL RICHARDS is Lecturer in Geography at the University of London and Coordinator of Research for the Environmental Review Unit of the IAI.

BEN WISNER has carried out research on economic development in Tanzania and Kenya and at the Institute of Development Studies, Sussex.

IAI

ENVIRONMENTAL RESEARCH REGISTER

The Environmental Review Unit of the IAI hopes to compile a register of individuals involved in environmental research about Africa who might be willing to collaborate in the future work of the Unit. The register will later be available in published form.

INSTITUTION:......................

Address

SURNAME:............................
(Capitals)
First names:........................

Position:

RESEARCH INTERESTS: (Please specify both subject and geographical region, e.g. 'Arid zone geomorphology, Northern Mali')	
RESEARCH IN PROGRESS:	Date begun: Projected completion date:
RELEVANT PUBLICATIONS (Please continue overleaf if necessary)	

Thank you very much indeed for your cooperation. Date:...................

Please send the completed questionnaire to: The Environmental Review Unit,
International African Institute,
210 High Holborn, London WC1V 7BW.

Future reports planned to follow the present volume include:

Rehab: the Continuing Famine in Ethiopia
Food Supply in Ethiopia
Food Supply in Nigeria
Land Use and Development
African Drought: an Integrated View.

The IAI will also compile EMERGENCY REPORTS as and when the need arises, designed to be of use to relief and development agencies as well as being of general interest.

Entries are being collected for an ENVIRONMENTAL RESEARCH REGISTER. Please complete the questionnaire in this volume, for the inclusion of your own name and research.

STANDING ORDER

FOR PUBLICATIONS OF THE ENVIRONMENTAL REVIEW UNIT

NAME : ....

ADDRESS: ....

....

.... Are you a member of IAI? ....*

Please send me .....copy/ies of each Special Report on the African Environment
.....copy/ies of any other Environmental Review Unit publication

In English
French
(Tick language preferred)

An invoice, covering postage and packing as well as the basic price, will be enclosed with the publication. The order may be cancelled at any time, provided one month's notice is given.

*NB: Membership of the Institute entitles the member to a 10 to 50% discount on IAI publications. See details of membership on back cover.

Signature: .... Date: ....

Please send completed forms to: The International African Institute,
210 High Holborn, London WC1V 7BW

PREFACE

The IAI's series of *African Environment Special Reports* is published in association with the quarterly bulletin *African Environment*, published by the Environment Training Programme, Dakar. Each organisation is responsible for the editorship and production of its own publications but there is close coordination of the materials included. It is intended that the Special Reports and the quarterly bulletin should supplement each other, and provide an effective forum for the discussion of environmental problems in Africa. The IAI, responsible since 1926 for research and publications on the societies, cultures and peoples of Africa, has recently established a new programme of applied research and documentation. The programme comprises an Environmental Review Unit and a Language and Education Unit, staffed by a small international research team. Among the first research fellows the team includes young scholars from Ethiopia, Morocco, Nigeria, Sierra Leone, Yugoslavia and the UK.

The Environmental Review Unit, which has published this report, could not, however, have undertaken its programme without the support of a number of donors, including the Leverhulme Trust Fund and Christian Aid and several African and European Governments. Further support is needed for specific research and documentation projects and for subsequent publications, and for the provision of research fellowships for scholars and scientists from African and non-African countries, enabling them to join the IAI research team. Those with research and practical experience would also be making a valuable contribution by sending offprints for the documentation centre and articles and monographs for consideration in the publications programme. Prospective contributors should refer to the section on Information for Contributors on page 117.

It should be emphasised that the IAI is both international and non-political in character, and that its publications provide a forum for scientific studies by writers of varying nationalities and persuasions. The IAI is a fully independent organisation, managed by an Executive Council comprising individual scholars and scientists from 13 African and 8 non-African countries (all of whom give their services to the IAI on an entirely honorary basis).

There are at present approx. 2,500 individual and corporate members, and no better support can be shown for the work of the IAI than by maintaining or taking out membership. The advantages of membership are being increased from 1976, the fiftieth anniversary of the IAI (see back cover of this publication), and it is hoped also that members and member institutions will participate as fully as possible in the research and publications programme of the IAI.

DAVID DALBY
Director
International African Institute

INTRODUCTION

The world over there is a growing concern for, and interest in, problems of the environment. Nowhere is this interest and concern more sharply focused than in Africa where five or six years of climatic abnormality have drawn attention to the especially delicate balance between man and nature which occurs in situations of harsh social inequality and economic deprivation. So severe in fact, has been the plight of drought victims in the Sahel, Ethiopia and Somalia that the normally leathery consciences of the richer nations of the world have been singed - seared would be too strong a word - and quite unusually grand international relief operations have resulted. Yet in many cases, this relief aid has raised more problems than it has solved, for, where a complete way of life has been dislocated if not totally destroyed, it takes more than free food and medicine to repair the damage done. It would perhaps be true to say therefore that, if drought in Africa has produced a heightened awareness of the role of environmental factors in the total development process, then experience of drought relief has given new emphasis to the fact that environmental management is as much a problem of political economy as it is of technological development.

Even if we dismiss sterile arguments about whether social science is sub or super science, it must be admitted that there are differences of method and language separating the social scientists on the one hand from the natural scientists and technologists on the other. There is a further and even more serious communications gap separating development theorists and high-level decision makers from the people - factory workers, farmers, administrators, doctors, teachers - who bear the day-to-day brunt of the development process at grass roots. Both these gaps must be bridged if satisfactorily integrated planning for environmental management is to be undertaken. The Environmental Review Unit of the International African Institute is trying to assist this process of communication in a number of ways - through its register of environmental research, through miscellaneous bibliographical essays aimed at the busy field project worker[1], through research into the ethnolinguistics of environment in Africa, and through a publications programme involving the bulletin *African Environment* (published by the Environment Training Programme UNEP-IDEP-SIDA in Dakar, in association with the IAI), a series of *Emergency Reports*[2], and finally a regular series of research papers and collected essays on specific themes to be known as *Special Reports* (published by IAI in association with the Environment Programme Training, Dakar). The present volume is the first of these *Special Reports*. Others planned or in preparation include studies of drought, food supply, the ecology of disease, water-supply problems, land use systems, and African 'traditional' environmental science. The series

[1]OXBY, C. 1975 *Pastoral Nomads and Development/Les Pasteurs Nomades face au Développement*. A select annotated bibliography with special reference to the Sahel, with an analytical introduction in English and French. IAI, London. 35pp.

[2]LEWIS, I.M. (ed.) 1975 *Abaar: the Somali Drought*. IAI, London. 44pp.

will complement *African Environment* to the extent that any comment on, or discussion of, controversial issues raised in the *Special Reports* can be carried on in the pages of the bulletin, and issues of especial interest to readers of *African Environment* might well blossom into complete *Special Reports* if sufficient material were forthcoming.

The present volume is a general introductory conspectus of environmental problems in Africa. It does not pretend to be comprehensive in coverage of topics nor even adequately representative of current research on the African environment. In a volume that has taken little more than six months from conception to printing we have had to depend very much on people being in a position to supply 'instant copy' and I would like to acknowledge the enthusiasm and hard work of all contributors in responding so quickly and generously to our initial request for material. Nevertheless, despite this constraint of time, we have tried to cover as wide a range of topics as possible and it is probably true to say most of the obviously important issues receive some attention. We have also tried to maintain some sort of balance between African-based and 'external' researchers, between workers with a highly technical 'natural science' approach and those with a social science orientation, and between francophone and anglophone contributions (future contributions in any other *lingua franca* would be welcome)[1]. Few contributors have risked defining environment - nor do I propose to grasp the nettle here. It is quite clear that it is a term with a wide range of meanings - for some it is the ecosystem, for others the natural resource base, and for others again the total social and material matrix of life. Perhaps the most thought-provoking idea of all is that just as natural processes such as lack of rainfall affect social structures, so social processes such as economic 'development' can affect natural systems, 'causing' famine and soil erosion for example. This notion, which crops up in various ways in several of the papers should make us think again about the term *natural* disaster. For in a continent where international ties of dependency, massive international labour migration and multinational companies prevail and in a world where growth does not necessarily mean development, and development does not necessarily bring enrichment or an increase in personal happiness, the ultimate cause of environmental problems may well be traceable to the structural imbalances between rich and poor countries and we would be right to replace the term *natural* disaster by the more appropriate term *social* or *political* disaster.

One final point is in order. Most of the papers in this volume are general surveys. That by Dissou is an exception, and by conventional reckoning it is not about the environment at all, but the importance of its approach cannot be overemphasised. It is a detailed case study of a small scale integrated development project for a rural area in Dahomey. As such it symbolises the greatest needs of the greater part of Africa. All the sophisticated techniques of environmental management from satellites downwards will prove pointless if the lessons of grassroots development schemes such as this are not learned properly.

Here then is a set of essays on the problems of the African environment - offered as a stimulus to discussion. I am sure that all the contributors would share with me the hope that the volume will help to provoke - perhaps as much by acts of omission as of commission - new and more deeply probing research into the practical realities of the greatest of all environmental problems - hunger and poverty.

PAUL RICHARDS

[1] The present volume is being published simultaneously in French and English. The original language of each paper is indicated in a footnote.

AFRICAN ENVIRONMENT
ENVIRONNEMENT AFRICAIN

A bilingual quarterly bulletin
published by
the Environment Training Programme, Dakar
in association with IAI.

Vol. 1.1 includes:
I. Sachs: Environment and Styles of Development.
G. Benneh: The Ecology of Peasant Farming Systems in Ghana.

Vol. 1.2 includes:
C. Toupet: Desertification south of the Sahara.
A Bibliography on the Famines and Drought of the Sahel.

1000 CFA F. (£2) per year.
600 CFA F. (£1.20) students.

Subscriptions should be sent to:
Environment Training Programme
B.P. 3370, Dakar, Senegal.

REFLECTIONS ON THE FUTURE OF THE SEMI-ARID REGIONS*

JEAN DRESCH
University of Paris

The semi-arid regions are transitional between the arid and humid zones. The consequences of this are none the less clear. The semi-arid regions may be occupied by populations adapted to either drier or wetter conditions, who are capable nevertheless of using the strictly limited resources, permanently threatened with insufficiency due to the irregularity of the bioclimatic conditions and the rapid and brutal effects of human actions. But, difficult as they are to define precisely - tropical or sub-tropical, continental or coastal - the semi-arid regions extend over considerable areas of our continents. At a time when the world seems too small, they pose difficult problems for development. Years of drought in the tropical semi-arid regions have resulted in scarcities, famines, the loss of the herds that were the people's capital, distress, and migration towards the less inhospitable surrounding areas. The affected countries, international organisations, and the countries of the world in general have had to intervene, albeit late in the day. But aid and charity do not bring about lasting solutions. Innumerable investigations have ended in very varied and contradictory programmes for action. How then can one form an opinion of what should be done?

I The Irregularities of Natural Conditions and their Consequences

Slender water resources are the main characteristic of the semi-arid regions. Delimiting the area by reference to average rainfall figures or extremes is of little use on the whole owing to the variations in the figures put forward. Precipitation is generally sufficient both to allow the development of pastures more prolific than in the 'desert' and to allow what is called 'dry-land' or 'rain-fed' agriculture, because it is possible without irrigation. The importance of the rainfall depends on whether it occurs in winter, the cooler season, when rain is more useful - that is, in so called sub-tropical, Mediterranean regions, where it amounts to between 150 and 3/400 mm - or whether it occurs in summer, the hot season - that is, in tropical, continental areas, where its usefulness is reduced by high evapotranspiration, and it amounts to between 150 and 5/600 mm. Agriculture sustained by rain is only possible if the rains are sufficient - 2/300 mm in the case of winter rains, more than 350 mm in the case of summer rains - begin early enough, and end sufficiently late to allow a long enough growing period according to the needs of different kinds of cereal crops. Animals are relatively less demanding. They can live in those areas of what is often called steppe or pseudo-steppe vegetation - open grassland and woodland - where the vegetation is specifically adapted, and renewed during the wet season, in accordance with the regional and local ecological and edaphic conditions. No less variable and strictly limited are the water resources, both in quantity and in quality. They may consist of

*French original

rainy season pools, wadis or swamps, flooded for a long or short period more or less frequently, permanent allogenic streams, areas of surface water or aquifers of varying importance according to the geological structure and the surface soil formations.

However variable these resources, they are always very limited. They can only be used, only be renewed, to the extent that an equilibrium is established and maintained between their own dynamics and the concentration, techniques and methods of production of those who use them - whether pastoralists with their camels, cattle, sheep and goats, or agriculturalists, or people who practise both. They are sometimes people adapted to the rhythms of the steppe who confine themselves to using this region alone. But often the steppe is used only as a supplementary, marginal zone. The nomad of the 'desert' takes his herds there in the dry season and stays if circumstances oblige him to settle. The farmer considers it an area for eventual colonisation if bioclimatic conditions are favourable. So both are involved in their own fashion in the use and renewal of the resources and adapt themselves to them. Adaptation depends on movement: the movement of the migrant pastoralist from the 'desert' to the steppe or further into the steppe as pasture is renewed; the movement of the farmer who extends, reduces or changes the land he clears each year according to the varying amount of rainfall.

Movement increases in fact with irregularities in climate, since, as is well known, the drier the climate, the more irregular the rainfall from year to year and the less significant the rainfall mean values. The variation from year to year is greater than 25%, often much greater. The renewal of resources is consequently a matter of chance and their use by man uncertain. Experience shows that pastoralists and farmers have in general been able to adapt themselves to the bad years, the years of badly distributed and insufficient rainfall. The former have their reserves of large and small stock which can, if necessary, be moved further than usual to pasture in less unfavourable regions. The latter, notably in the tropical Sahel region, have their grain store, an essential and characteristic element of this habitat

But these safeguards have their limitations. Thus they are only effective in the case of short interruptions in the equilibrium. If there is too long a succession of dry years there may be famine, often followed by epidemics. In any case this will bring about the ruin of the pastoralist, in particular, because it is difficult for him to reconstitute decimated herds. In all the semi-arid regions the memory is handed down of terrible periods when camps and villages were emptied and abandoned for a long time. But, normally, better years brought life back again. Life then went on as before, provided that, between one drought crisis and another the population did not increase, that the needs of the population were very simple, and that the activities of each generation had as their principal aim and effect the survival and continuity of the group. The rock paintings of the cattle herders of the neolithic Sahara, 5000 years ago, evoke the methods of production and way of life of the present day nomads of the Sahel. But in the 20th century it is becoming more and more difficult to think in terms of human groups whose only end in life is to survive, conserving their methods of production and their standard of living in a state of isolation - not indeed in comparison with neighbouring regions, whether drier or wetter, but within a world which has to search and prospect for all useful resources and to satisfy ever-increasing needs. Our technological civilisation has set more and more rigorous standards of international fellowship, including the moral obligation, in theory, to stop human communities from dying and to satisfy new needs that are considered to represent a particular conception of human dignity.

II The Problems

Whatever the continuing progress of science, it has not so far been able to modify semi-arid climates. Rain-making experiments only have a very limited effect and are costly. We can do better by increasing and making better use of our water resources. We cannot at present create them, except by desalination of salt water by expensive processes, used so far only on sea coasts to obtain drinking or industrial water. If only it were possible to foresee sequences of dry and wet years! Since the time of ancient Egypt men have imagined and hoped for such prescience, but all attempts have failed. To the extent that reliable statistics exist for sufficiently long periods, which is rare in the case of under-developed countries, careful examination of these gives no grounds for forming any conclusion about a periodic pattern of sequences, and in any case the sequences are not the same from one region to another. In the West African Sahel, series of dry years are recorded for 1900-03, 1909-16 (especially 1913-14), 1931, 1940-44, and then 1968-73. These do not correspond with the series of dry years on the sub-tropical, Mediterranean edge of the Sahara which are for the most part the reverse. They have not been of the same severity in the whole of the Sahel, except for the most recent drought, although it is true that documentation for this drought has been less incomplete.

A relationship has been sought between these series and other phenomena, such as sun spot cycles. A possible explanation postulates a relationship with the general atmospheric circulation and particularly with the circulation over the South Atlantic: when it is rapid, this circulation pushes the tropical high pressure zone back towards the north and consequently increases the force of the monsoon in the northern hemisphere; when it is slow, as was the case between 1968 and 1973, it reduces the monsoon's force. One can thus try to explain the oscillations or changes in the climate on different time scales: those changes which have brought about, amongst other phenomena, the fluctuations over the last century in the level of Lake Chad and a tendency for the average rainfall at Saint-Louis to decrease; those changes evidenced, in the same region, by the advance towards the north of agricultural exploitation of land by villagers (Gangara) who, between the 11th and the 15th centuries were able to cultivate millet more than 200 km to the north of the present zone of cultivation; and those changes which explain the alternation since 40,000 B.P. of dry periods and pluvials during which lakes and ponds occurred in the present Sahara desert and pastures were sufficiently abundant to allow cattle herding.

If it is impossible to explain definitively the different modifications of the climate in inter-tropical zones, modifications which do not even appear to correspond on the two sides, north and south, of the areas of tropical high pressure, it does at least appear to be possible to distinguish trends in the evolution of the general atmospheric circulation, related to, amongst other phenomena, changes in precipitation in the tropical zone or glacial periods in the temperate and cold zones. This is a preliminary step towards forecasting for the medium term. This forecast may be unfavourable for the Sahel countries if, for example, it is true that in a glacial period the acceleration and movement towards the south of the atmospheric circulation in the northern hemisphere could have as its consequence the movement of the Sahara towards the south and the prolonged desertification of the Sahel, as happened at least once in the course of the recent Pleistocene period. In this event the problem of human occupation of the Sahel would be simple. There would be no other solution except to provide for evacuation. But it is right to be cautious: we do not know anything of the chronology and changes in the climate in the southern Sahara and the Sahel, during the Quaternary period, earlier than 40,000 years ago. We can guess that recent phases of the Würm glacial in the temperate zone were dry periods in the Sahel but we cannot extend these conclusions to earlier phases and periods. In addition in the Sahel the summer of 1974 was fairly wet and the vegetation, degraded by 5 to 6 years of drought, has recovered with a surprising vitality.

But irregularities from year to year, and oscillations or changes in the climate are not the only problems of the semi-arid regions. At present they are not even the most serious. Socio-economic changes are much more important. During the precolonial period, pastoralists and farmers in turn established their supremacy over the semi-arid regions that border the Sahara to the north and to the south. The history of the Middle East and of Arabic North Africa, as of the West African Sahel, has often been summed up as a continuing conflict between nomadic and settled peoples since both draw their resources, at least in part, from the semi-arid stepp A correlation has even been suggested between the advance of the nomads towards the margins of the desert and a dry climatic oscillation, readily seized upon as its cause. It is clear that such a bioclimatic causation is a simplistic interpretation of the train of events and that the reasons for the Arab conquest of the 7th century, the penetration of the nomadic Berber (Zenata or Sanhadja Almoravids) or of the Hilalian nomadic Arabs (11th cent into North Africa are much more political than climatic. Nor can the end o the great Western Sudanic empires of the 11th to 15th centuries - Ghana, Ma Songhaï - be explained by a deterioration in the climate from the beginning the 15th century, although this may have played some part in the process.

Equally the advance of the nomads towards the two edges of the Sahara has not had the grave consequences for the natural environment that are fre quently attributed to it by Arab and, even more, Western historians. The Hilalian Arabs are accused of having systematically cut down the fruit tree of North Africa; their sheep, and even more their goats, are said to have hastened the destruction of the woodlands. The same accusations have been made against the Arabs of the Middle East. But even if damage caused by the flocks cannot be denied, though it has been exaggerated, clearance by farme and their various uses of timber have been much more damaging, especially since their population density is so much greater. One can argue at length about how far the nomadic pastoralist contributes to the degradation of the vegetation cover. Everything depends on the relationship between the pastu resources and the grazing load put upon them. The former vary according to the rainfall, the latter varies according to demographic or economic condit In a dry period, the pastoralist cuts down the trees to provide leaves for fodder. He increases his flock excessively when it is profitable. However this may be, the pastoralist can be considered to be a sufficiently good conserver of the vegetation cover in 'normal' conditions, as long as the ar is not degraded by overgrazing arising from causes other than insufficient rainfall. At least it is certain that in the West African Sahel the suprem of the pastoral nomads between the 15th and the end of the 19th century contributed to the low human population density, since their birth-rate is generally lower than that of settled people, and low animal population dens since the commercial use for animals was limited.

The colonial occupation has upset this equilibrium, as much on the Mediterranean as on the Sahelian edge of the desert. In the Maghreb, the extension of agricultural settlement first on the richest soils and then into the semi-arid regions, thanks to the adoption of dry-farming methods for cereal cultivation; the land hunger of the local population and the progress of clearing the land, even on hillslopes, at the expense of pasture and of forests growing at their bioclimatic limits; the sedentarisation of nomads and semi-nomads and the decline in the rearing of sheep, which were not valued by the colonisers; and the more and more rapid increase in the population have resulted in the destruction or driving back of the forest, the climax vegetation, already more or less deteriorated, and the dangerous acceleration of soil erosion.

In so sensitive a morphogenetic area, where slopes are often steep, low resistant rocks are often present and climatic changes extreme, all types of soil erosion are to be found side by side: all degrees of mass movement of soil, gullying, sheet erosion, wind erosion. Erosion is progressing at such a rate that the results often appear to be irreversible, the struggle against them is technically difficult, and in any case slow and very costly. Retaining barriers are silted up before they have been fully effective. Very disquieting estimates have been made of the losses of arable land arising from this: 40,000 ha. of cultivated land lost annually, out of 5 million ha. threatened, in non-Saharan Algeria. These estimates are exaggerated, but the specific degradation may reach nearly 2000 tonnes per km^2 on average (Wadi Isser in central Algeria), corresponding to the removal of a cover of more than 3 mm on average per annum in the basin (Wadi Fodda in the Ouarsenis). In different socio-political conditions, a comparable situation can be observed in the Arab countries of the Middle East.

In the Sahel, nearly everywhere a much flatter area, where the bioclimatic characteristics of the environment are very different, the socio-economic conditions are no less different. The Sahel was of no interest to the coloniser. It was further to the south that commercial cultivation became important - groundnuts in the regions where the rainfall exceeds 4/500 mm, and cotton in regions where rainfall exceeds 6/800 mm. The Sahel, or at least the driest part of the Sahel, thus became a provider of millet or of meat for the countries orientated towards export monoculture, or it was simply ignored because it did not fit into the colonial economy. In these conditions, where was the point in equipping it with roads, schools, hospitals? Nevertheless, colonisation did bring about profound changes even in the Sahel. Nomadic society has lost part of its influence, political as well as economic. The caravans are no longer able to perform their international function; the products of the desert - animals , salt - have lost some of their value as a means of exchange; the traditional, very hierarchical, social structures have disintegrated. Slaves and captives (*iklan, bolla, bugaje*) have gradually been freed, their numbers are increasing faster than those of higher castes, and they are migrating towards the Sahel and establishing themselves there. This agro-pastoral colonisation has come into collision with the expansion of the Fula. The latter, especially in wet years, now drive their cattle more freely towards the north, 200 km beyond their traditional migratory limits in Mauretania and Niger. Thus colonisation from the 'desert' can be set against agricultural colonisation from the wetter Sahel or from the Sudan savanna zone especially, of course, in sequences of wet years. This equally limits the movements of pastoralists, keeping them from the pools or river-banks, now devoted to flood-retreat or rice cultivation, thus provoking conflicts. In these conditions, the population increases, resources are more fully used, pasture is overgrazed, particularly around pools and wells, the lighter soils, often on old dunes, suffer erosion by removal of the cover or the formation of ravines, or the action of the wind. It has been estimated that erosion of Sahelian soils is 10 to 20 tonnes per ha. per annum.

Thus, on the edges of the Sahara, both north and south, settlement has in general had evil consequences: the natural resources have been heedlessly overexploited. They have hardly been integrated at all into a 'developed economy, any more than the people have been. In consequence, the uncertainties and the threat of disequilibrium in these marginal regions have been aggravated. Independent, but 'under-developed', states have not yet been able to find a solution.

III What are the Solutions?

In the steppes of the Maghreb, nomadism or semi-nomadism has disappeared (Tunisia) or is diminishing in extent and losing its traditional socio-political characteristics. The extension and mechanisation of cereal cultiva-tion reduce the extent of the grazing areas. In the low and the high Tunisia steppes plantations of olives, almonds and apricots have the same results. In Morocco, the extension of the area under irrigation in the western foot-hills of the Middle Atlas and in the Haouz of Marrakesh prevent the semi-noma or transhumant mountain pastoralists from using their customary traditional winter pastures. The Algerian government is extending the agrarian revolutio into the steppes and is trying to modernise sheep rearing. It has undertaken the planting of a long forest strip intended in theory to stop the spread of the desert. It is, in fact, true that climatic aridity is not extending northwards. At least the forest barrier indicates a desire to stop man-made extensions of barren land. Equally the administrations of the three states of the Maghreb have extended considerably the areas of soil conservation and reafforestation. It is less difficult for independent states than for coloni powers to control the inevitable conflicts between forest rangers and traditional users of the forests and steppes. After all, as in the Middle East, if the deserts gradually become 'deserted' because the nomads who occupied them settle down more and more, moving towards the margins of the desert, generally under pressure from governments, it would seem that these semi-arid margins could be organised and integrated into the national economies of developing states.

It is much more difficult to speculate on the future of the West Afric Sahel. The first years of independence coincided with a sequence of wet year i.e. until 1968. The cattle herdsmen - 'slave' Tuareg starting to settle, or Fula - allowed their herds to multiply; the farmers extended their areas of cultivation and tried to push their frontier of settlement towards the north, to the point where the state authorities had to intervene to limit it and to control relations between pastoralists and agriculturalists. When a series o dry years succeeded each other, after 1968, the pressure of pastoralism becam clear, particularly around wells, more and more of which had been dug, oft with the best of intentions but without the organisation of any discipline in their use. The cattle were clearly the victims most threatened by the degrad tion of the pastures; goats were more resilient. The Fula, who are excellent stock rearers, often had the wisdom to migrate early enough to the savanna zo where they could hope to find more favourable conditions. The Tuareg hesitat longer and found it more difficult to maintain discipline in their herding. Herds were decimated, and sometimes completely destroyed. In general, they could not even be slaughtered and sold, not because of the 'cattle worship' often attributed to the Fula, but because trade was badly organised and there was no method of rapid disposal of stock. The decimation of the cattle herds must be attributed not only to the drought but also to under-development.

When rainfall became sufficient again, in 1947, pastures often revived more quickly than expected,and herdsmen have mostly tried to return to their habitual grazing grounds in so far as they could restore their stock. But what lesson should one draw from a drama that has brought tardy sympathy from the part of the world which is not hungry? A number of measures could be taken to develop the pastoral economy: to improve the quality of pasture and of fallow land by enriching them, although the difficulties here are numerous; to combine nomadic pasturing in dry areas with periodic stops in irrigated land, as in Soviet Central Asia; to improve breed quality; to increase watering places and above all to take care to maintain a constant equilibrium between stocking and pastoral resources, by regulating access to watering places and to pastures; to plan ranches where these measures could be put into practice; to foster the profitable marketing of cattle and their produce by constructing abattoirs, dairies, cheese factories, by organising agro-pastoral undertakings or by making the Sahel specialise in the breeding of young stock for supply to the wetter savanna regions for fattening; in fact to improve the commercial network and organisation of the market in general which are at present too often unfavourable to the producer.

Measures could also be taken to develop agriculture. To combat the irregularity of rainfall, it is clear that it would be desirable to develop irrigation, using all water resources, through the construction of dams on the rivers (Niger, Senegal), lesser dams in topographical depressions to collect stream water, and the sinking of deep wells to allow the use of underground water supplies where these exist. Such construction work needs to be accompanied by control of the irrigated area in such a way as to avoid water wastage as far as possible and to ensure drainage. A programme of malaria, bilharzia and even onchocercosis control would have to be undertaken. Flood-retreat agriculture can be combined with irrigated cultivation. In irrigated as in rain-fed cultivation food or industrial crops can be raised in association with animal fodder, cereals with leguminous crops, and stock rearing can be combining with agriculture. Cultivation by animal power, which is generally preferable to mechanisation, enables a farmer to extend the area of cultivation, and animal manure can be used with chemical fertilisers, thus reducing the amount of fallow land needed. The general result is the emergence of agro-pastoral and agro-industrial complexes. The selection of suitable crop varieties, the improvement of techniques, the introduction of a greater range of crops into the cultivation system, the adoption of better methods for storing the harvest and for marketing it - all these are the most common ideas put forward and tried out.

But it is all too easy to suggest improved techniques and even to conduct experiments. Experience shows that it is more difficult to integrate them into the economy of an under-developed country and to make them acceptable to societies that remain faithful to traditional methods of production. Such improvements in effect assume a Sahel equipped with a communications network and with administrative, technical, commercial, public health and educational services; they assume a personnel combining technical qualifications with great devotion to duty and feeling for the pastoralists and the farmers whom they advise and instruct: and they assume, finally, a profounc transformation, necessarily slow, of the traditional social structure such that ethnic rivalries would be reduced, reforms would be accepted, and those concerned would participate actively in the reforms. For a programme of such magnitude and complexity, the states, amongst the poorest in Africa and in the world in general, would have to provide for considerable investments which would not pay off for a long time.

It is possible therefore to maintain a population in the Sahel and to improve their living conditions. But one should not ignore certain facts. There is no question at present of any profound modification of the natural environment through artificial rain-making, desalination of water, or diverting the course of major rivers. Even if it were possible to use the natural resources better - to use them optimally - they would nonetheless be more limited than in wetter regions. For a long time, the problem will continue to be to get a balance between resources and their exploitation by man, through limiting the stock using the pasture, the areas brought under cultivation and the size of the population itself (by birth control or by organis emigration). In the immediate future, limited human occupation can be restored and maintained only if the needs of the people are also limited and if the improvement in the standard of living is maintained at a very steady ra In 1968 the income per head was $90 in Mali, $40 in Upper Volta and in every case less than $160 in countries with similar conditions. One knows what these averages mean. Is it right to maintain pastoral nomadism if it does not allow more than this? Should the cultivation of millet with its low yield and low value be maintained? In theory one can ask these questions. But can one on the contrary imagine that there is no other solution but to evacuate the Sahel? Would the neighbouring wetter savanna areas be able to receive the Sahelians and ensure them a higher standard of living by integrat ing them economically, socially and indeed politically into a nation state It is clear that even if these solutions might be envisaged for the middle term, they cannot be envisaged in the short term. A careful improvement of Sahelian territories taking into account all the physical and human factors is the only possible solution at the present time. Even so, it will not be possible to remove entirely anxiety about drought, the threat of famine or the poverty of the inhabitants.

ENVIRONMENTAL MONITORING NEEDS IN AFRICA*

BENGT LUNDHOLM
Secretariat for International Ecology, Sweden

The word 'monitoring' has different meanings and it will be defined here as regular synoptic observations in a space-time series, designed to give information about the environment (global, regional and national, or local) so that past and existing states can be assessed and future trends predicted for any environmental feature which may be important to man. This definition is from the report of a task force of the Man and the Biosphere (MAB) Programme dealing with monitoring research (UNESCO, 1974).

From this definition it is clear that environmental monitoring has to be problem-oriented. These problems could be divided into three main categories:

1. Potentially adverse climatic changes resulting from human activities.

2. Potentially adverse changes in biota and man from contamination by toxic substances.

3. Potentially adverse changes in biological productivity caused by improper land use.

Those environmental parameters that describe and quantify these three problem areas can be classified under the following main headings:

1 a. Physical and chemical data from the atmosphere pertinent to climatic change potential.

1 b. Physical and chemical data from air, water, soils, and biota pertinent to human health and welfare.

2 a. Physical, chemical, and biological data reflecting the state of human health.

2 b. Biological data reflecting the performance of biological systems.

In order to measure these variables, special monitoring systems have to be set up. Recommendations for such system were made to the 1972 United Nations Conference on the Human Environment held in Stockholm. Responsible for the global monitoring activities is the UN Environmental

*English original

Programme (UNEP), and it has now been decided that a special Global Environmental Monitoring System (GEMS), should be established. An inter-governmental meeting on monitoring, convened by UNEP, was held in Nairobi during February 1974. The goals for GEMS were established as follows:

1. An expanded human health warning system.
2. An assessment of global atmospheric pollution and its impact on climate.
3. An assessment of the extent and distribution of contaminants in biological systems, particularly food chains.
4. An assessment of critical environmental problems relating to agriculture and to land and water use.
5. An assessment of the response of terrestrial ecosystems to pressures exerted on the environment.
6. An assessment of the state of ocean pollution and its impact on marine ecosystems.
7. An improved international system allowing the monitoring of factors necessary for the understanding and forecasting of disasters and for the implementation of an efficient warning system.

At the same meeting, a list of priority pollutants was accepted, starting with sulphur dioxide and radio-nuclides as first priority and with ozone, DDT, and other organo-chlorine compounds as second priority, followed by other groups of compounds as third, fourth, etc. priorities.

At the UN General Assembly at the end of 1974 a resolution was passed that GEMS should not only concentrate on priority pollutants but should devote half of its capacity to monitoring natural resources. This resolution was strongly supported by countries from the developing world.

This paper concentrates on aspects relating to natural resources and deals especially with the monitoring of dust and its importance in relation to arid and semi-arid ecosystems. Figure 1 shows a picture taken from the geosynchronous satellite SMS 1 on 8 July 1974. A large dust cloud that left Africa on 1 July is seen over the Atlantic. The dust in the cloud is at a height of 3000-4000 metres. Already in 1970 the Nimbus weather satellite had picked up dust clouds over the Atlantic. The consequences of this dust transport from Africa over the Atlantic have recently been analysed by Rapp (1974). The facts presented here are from this report.

That dust from Africa falls into the Atlantic has been known for a long time. Figure 2 is a map of dust recorded by ships at sea based on data collected before 1942 (Kuenen, 1950). In 1966, American scientists wanted to test clean air and as a test site they picked Barbados in the West Indies. But the air was not clean; it contained dust transported by easterly winds. In 1966, the dust content was 6 micrograms

Fig. 1. Dust cloud over the Atlantic. The cloud is marked with arrows. NASA-NOAA.

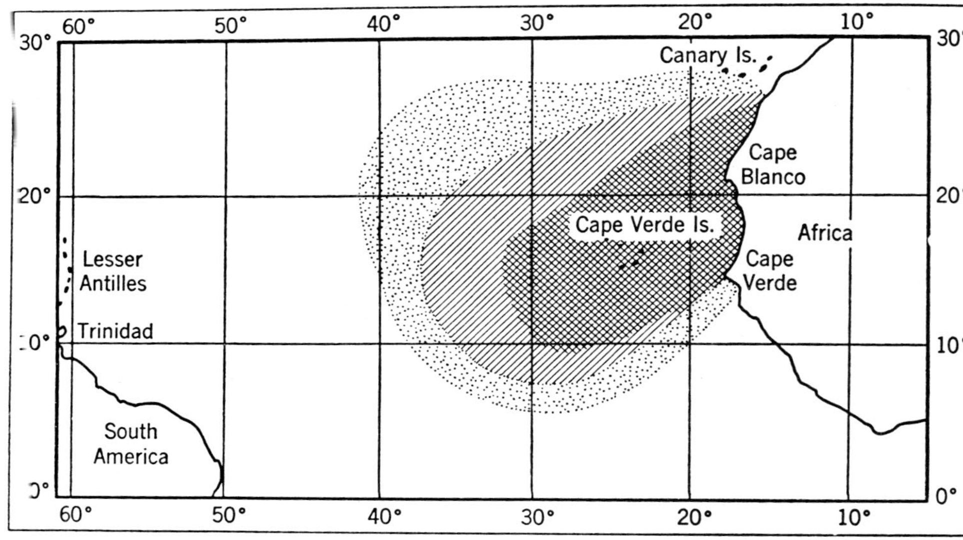

Fig. 2. Dustfall over the Atlantic. After Kuenen, 1950.

Fig. 3. Desert encroachment, El Hamma oasis, Tunisia. Two kinds of land degradation from wind erosion are visible: 1) land dunes advancing into the date palm grove in the centre of the picture, 2) deflation patches along cattle track in the right foreground. Photograph and interpretation from Anders Rapp.

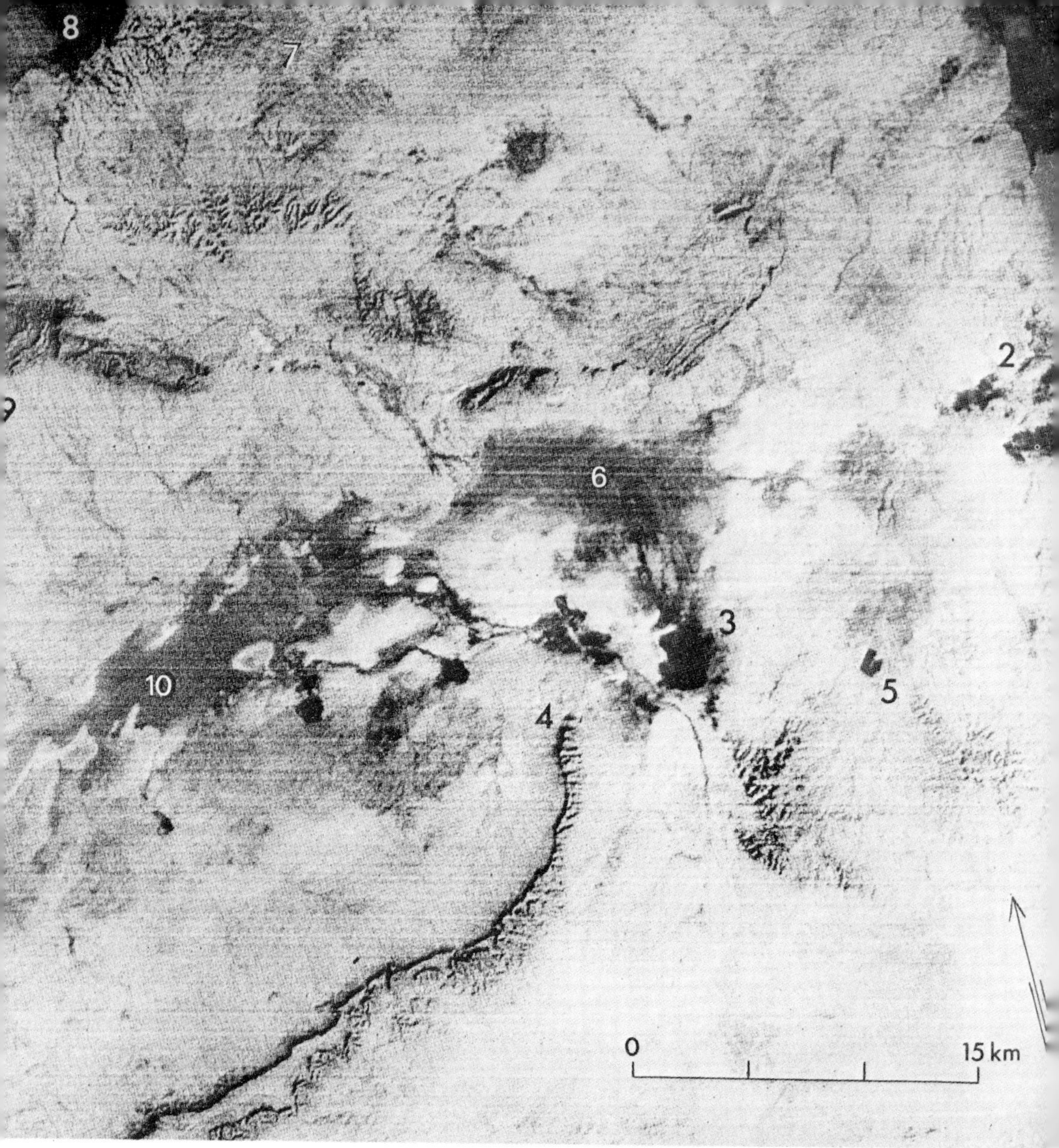

Fig. 4. Detail of ERTS-1 satellite image of El Hamma area. February 7, 1973 - Band 5. 3 = El Hamma oasis with dune in front and wind erosion scar on the left side. After Rapp, 1974.

per cubic metre. In 1967-68 it increased to 8 micrograms. After 1968 it increased rapidly to reach 15 micrograms in 1972 and 24 micrograms in the summer of 1973 (cf. Rapp, 1974). Here is the Sahel catastrophe! It is important to note, however, that the dust is not sand from the Sahara but African soil. During the summer months the dust is reddish-brown, probably coming from the northern margins of the Sahara; in winter the dust is black, coming from the burnt areas in the south. Deshler (1974) registered the extent of grass fires in the savannas of Africa along the southern side of the Sahara by using remote sensing and pictures from ERTS 1.

Rapp has pointed out that it is possible to study erosion phenomena by using ERTS 1 pictures, as is demonstrated in the following two figures. Figure 3 shows erosion of two types with loose sand dunes and deflation patches in Tunisia. Figure 4 is an ERTS 1 satellite photograph , processed in Sweden. The white ridge marked 3 corresponds to the dune on the previous figure. It has also been possible to trace the origin of specific dust storms from the Mojave Desert by using satellite pictures as shown by Bowden *et al.* (1974). Thus, it should not only be possible to record the amount of dust blown away but also to pinpoint the location of origin. MacLeod (1974) has pointed out that ERTS 1 photographs over the Sahel show thousands of square kilometres covered by moving sand during dust storms in 1973-74.

Prospero & Carlson (1972) have worked out the amount of dust blown from West Africa during the summer of 1969 to be about 60 million tons (see Fig.5). For comparison, it may be mentioned that the annual sediment transport in the Nile river has been estimated at 150-200 million tons at the inflow to Lake Nasser; thus they are of the same magnitude.

The conclusions in Rapp (1974) were:

> 'The types, extent, and rates of desertization processes have to be surveyed and monitored in the future as a basis for both short-term and long-term actions against desertization and its destructive effects.
>
> In agreement with many authors and reports we strongly recommend careful sampling surveys on the ground combined with extrapolation using aerial photographs and satellite data. To such surveying/monitoring transects have to be added pilot trial plots for the purpose of research and demonstrations of ecologically sound methods of land use in areas of potential or actual desertization.
>
> Recent discoveries show surprisingly large amounts and long transport distances of airborne soil dust from the Saharan and Sahelian zones into and over the Atlantic. This is probably the most destructive effect of desertization in a longer time perspective. We strongly recommend to UNEP and WMO to consider monitoring of airborne soil dust together with the use of satellite images, as an early warning system. Three suitable locations for ground stations would be:
>
> 1. The Cape Verde Islands (Sahelian west)
> 2. Jebel Marra Mountains, Sudan (Sahelian east)
> 3. Tenerife Island (northern Pre-Saharan belt of desertization).'

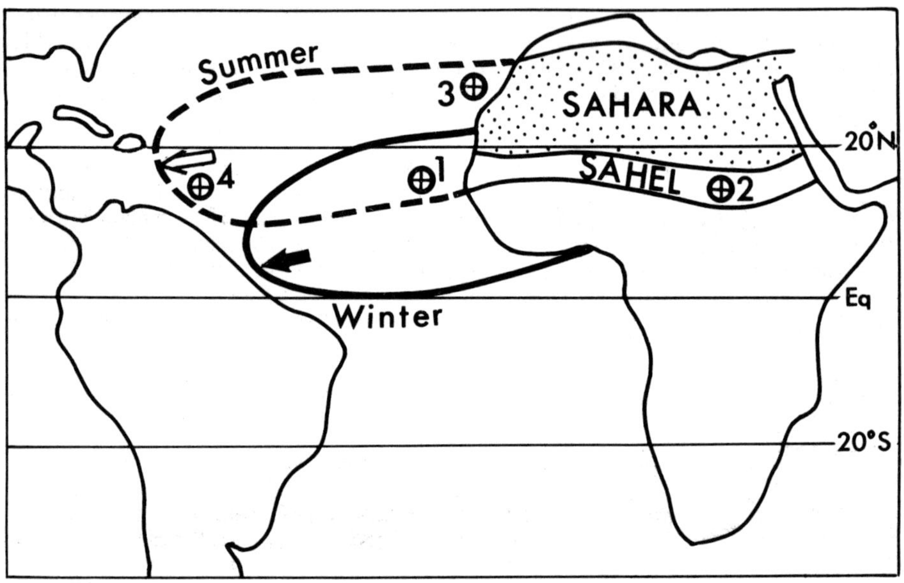

Fig. 5 Dust transport over the Atlantic.
1 = Tenerife. 2 - Jebel Marra Mountains, Sudan.
3 = Cape Verde Islands. 4 = Barbados. After Rapp, 1974.

The Secretariat for International Ecology, Sweden (SIES) is now trying to formulate pilot projects to explore these possibilities, especially the techniques for the measuring of dust.

If one tries to assess the effects of this dust transport in Africa one has to consider the global aspects first. It has been pointed out that the increased turbidity in the atmosphere may be responsible for the present cooling of the earth that began about 1945. It was recently reported, for instance, that the snow and ice cover increased by 12% during 1967-72 (Kukla and Kukla, 1974). Bryson and Wendland (1970) argued that increased turbidity was caused not only by industrial activities in temperate regions but also to a large extent by improper land use in tropical and subtropical regions. It is thus reasonable to assume that the dust catastrophe in Africa has to be studied when global climatic changes are considered. MacLeod (1974) has pointed out that the increased dust in the Sahel region may reduce the rainfall by suppressing the formation of rain-producing clouds and retarding the normal northward progress of the summer monsoon systems which bring rain to the Sahel.

The loss of topsoil is very important. It has to be stressed especially that, because of soil moisture movements in arid areas, the plant nutrients are concentrated in the surface layer. In this surface layer there are also micro-organisms responsible for nitrogen fixation. The erosion process results in transportation and sedimentation . The dry fallout over the Atlantic was shown in Fig. 2, and the sedimentation of nutrients over the Atlantic must be of importance for the productivity of the sea. This dry fallout must also be of importance to the terrestrial ecosystems. It is well-known that there is an accumulation of the dry sediment in vegetation, although very little is known about the importance of this redistribution of nutrients. It is also well-known that the fertility of the Nile Basin used to depend on the sediment load brought by the annual floods. Some have assumed that the fertility of the Nile Valley is bound to decrease as sediments accumulate in Lake Nasser. It has been claimed, however, that part of this fertility depends on nutrients blown from the neighbouring deserts. Verification of this and investigation of the importance and the consequences of sedimentation by wind-blown material must be regarded as a very urgent problem in Africa. It is of special importance in relation to livestock management. Here integrated ecosystem studies are needed to establish changes in productivity.

To find relevant measurable variables is the main difficulty with any kind of environmental monitoring. It must be realised that each measurement taken depends on the total situation in the environment. The same results from various environments may be evaluated quite differently. Of importance for each measurement are the end results of a long development. Thus, both the holistic situation and historic development have to be considered if one wants to assess the value of a single measurement. Only when one has long time-series of the physical and chemical variables is it possible to get a proper background for ecological studies.

A practical approach to the study of these monitoring problems in humid climates is to use small watersheds (SWS) as indicator areas for environmental changes. Since water is the main carrier of nutrients to and from the watershed, this transport can be measured. Small in

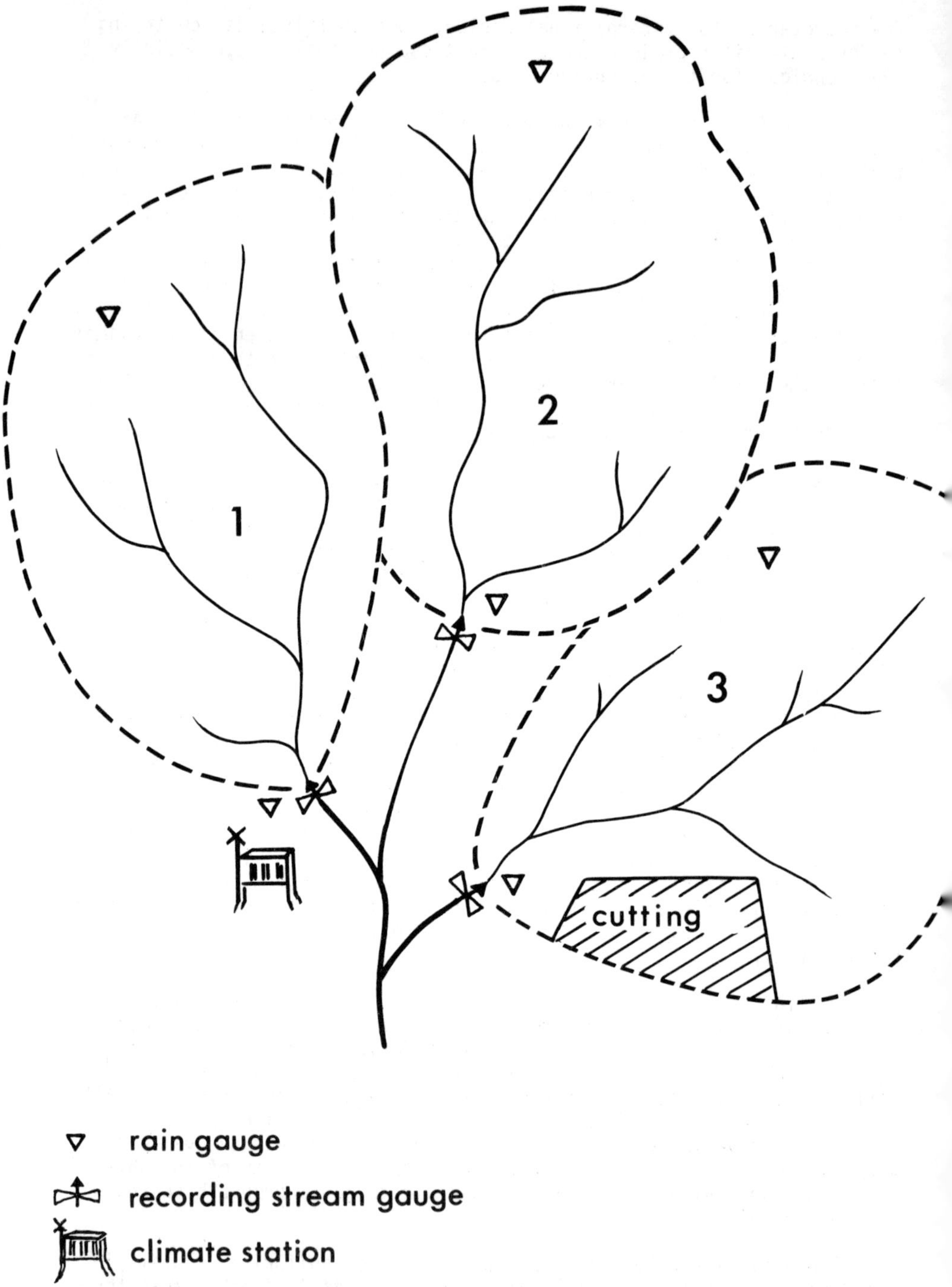

Fig. 6 Monitoring a cluster of small watersheds (SWS)

this context means a size of 0.2 - 2 km^2. In such areas it is possible to control both the input and the output of chemical and physical variables. A detailed study of the chemical balance is possible using well established methods.

Within each SWS the following variables should be observed:

1. The chemical input registered by monthly analyses of precipitation. Samples will contain rainwater (and snow) with its chemical content and also dry fallout. The following variables should be considered: SO_4, Cl, NO_3, NH_3, Na, K, Mg, Ca, alkalinity, pH, and electrical conductivity. These are routine measurements. In a more advanced stage, dry fallout should be measured separately as well as impaction and stem flow. It is important that both quantities and qualities are measured. To complete the picture, the standard meteorological measurements should be made for precipitation, temperature, wind direction and force, etc.

2. Output should be assessed by registering outflow water and analysing the same chemical components as mentioned under point 1. It is especially important to have a continuous registration of the water flow by using weirs.

3. A careful inventory of vegetation and soils as well as investigations of the history of vegetation and soil in the SWS should be made when the station is started. Inventories should be made at regular intervals to assess possible changes.

The program for such an SWS is quite simple, therefore, and the instrumentation required is inexpensive. Such a station could be run by technicians after a short training. It is possible, of course, to conduct much more sophisticated investigations in these indicator areas, such as measuring the chlorinated hydrocarbons and heavy metals.

To study the impact of human activities, clusters of small watersheds could be used in suitable areas (Fig. 6). In such a cluster of small watersheds only one climate station is needed, and the management of the instruments and the analysis work can be rationalised so that the cost would be much lower than the sum of a corresponding number of separate SWS. Even if differences in land use do not exist nor are expected to occur in an area where it is possible to set up a cluster of SWS, it might still be advantageous to use a cluster instead of a single SWS. From monitoring several SWS the results will have greater significance and will offer important possibilities for future experimentation with different land use mixes.

For the SWS so far outlined, rather simple variables are recorded, mostly water flow and chemical concentrations. The most important consequences of investigating these watersheds, when flow of matter is under control, is that they will be of great value for future ecological studies. At the present time it is difficult to find relevant variables for recording biological changes. There are suggestions for measuring biomass in different ways, using various indicator species, measuring the decomposing rate, etc. So far there are no international agreements on which parameters and which methods should be used. All these methods, however, have one thing in common, namely that they require background information on input and output. This information can only be assessed and properly evaluated when there are time series of sufficient length. By establishing SWS, such data could be gathered as background and reference material forecological studies. This is especially important in relation to studies of soils, where the adaptation to changes takes a long time.

A global network of SWS would be of importance for the Global Environmental Monitoring System (GEMS), which is now being set up by UNEP. The UNESCO Man and the Biosphere Programme (MAB) is also currently setting up biosphere reserves for ecological monitoring. In each of these reserves SWS ought to be established to provide background information on ecological changes. The establishment of SWS involves simple and inexpensive instrumentation. In temperate and cold areas electrical heating of the instruments might be required, considerably increasing costs. In tropical and subtropical climates, however, the costs can be kept very low. The chemical analyses are also simple and inexpensive. These conditions make the SWS especially well suited to the developing countries.

In dry areas the boundaries of the local ecosystems are more diffuse. The wind as a transport medium is dominant. The unit is not the watershed, but the airshed. The airshed has more diffuse boundaries and is also in general much larger than the watershed. Studies of the ecology of such areas are more difficult to make and so far have not been very well-developed. It is possible that the best results will be achieved by studies of transects through the air, which means studies over long distances. At every point in the transect both horizontal and vertical transport have to be followed.

There is a very close relationship between the use of natural renewable resources and pollution. The pollutant may decrease productivity by suppressing some vital processes; for instance, the acid rain over Europe may decrease productivity in forests and fresh waters. The effects of the pollutants - with so-called biological magnification, and their accumulation in the different compartments of the ecosystem - are well-known in humid areas.

In Fig. 7 one can see that the final sink for the pollutants is the sediment. The pollutants are trapped in the sediments and often sealed off. In certain cases the sediments are disturbed and the pollutants get into circulation again. Pollutants following this pattern include the metals mercury, lead and cadmium. In a dry area the situation is quite different (Fig. 8). The emissions are not washed out, but there is dry fallout and sedimentation. These dry sediments are easily disturbed, especially by wind erosion, and pollutants get into circulation again. The figure shows erosion caused by overgrazing. These pollutants are accumulated in the grass and then finally deposited in grazing animals.

This is a theoretical model and very few investigations have been carried out to verify the model. Some details, however, are well-known. It is known for instance that there is a dry fallout aroung industries emitting heavy metals. Well-established methods exist for the study of this phenomenon. It is even known that such pollutants can damage vegetation and can accumulate in grass and in grazing mammals. Thus, the processes are known but they have not been studied in depth for the dry ecosystem.

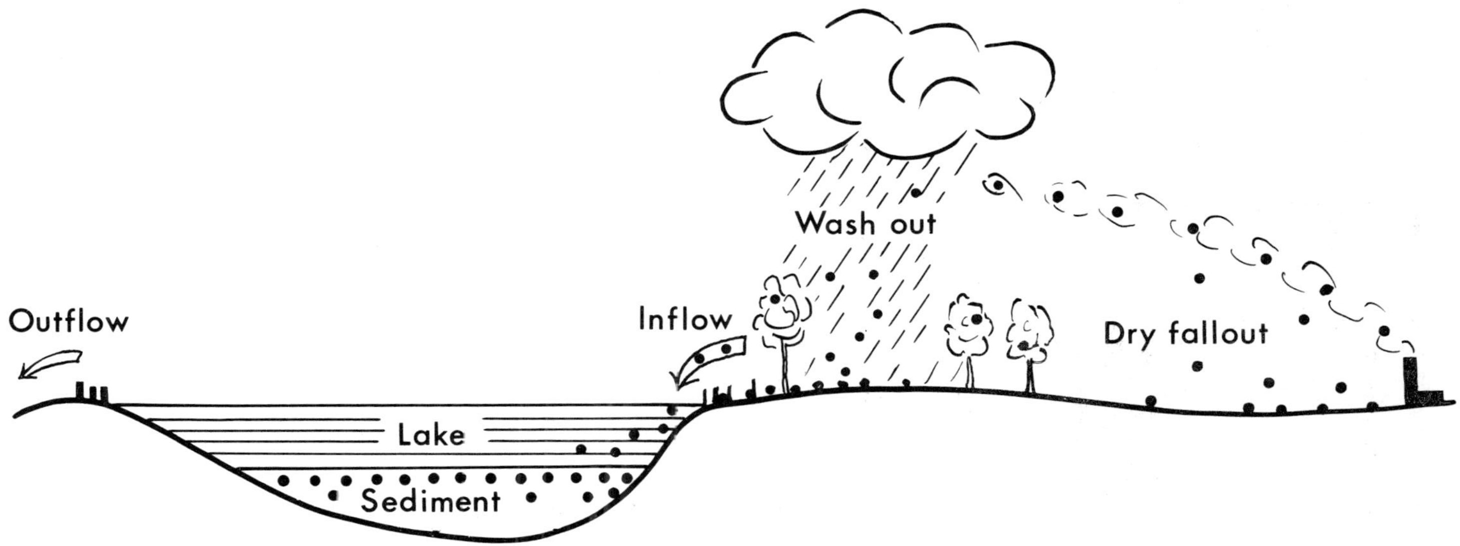

Fig. 7 **EMISSIONS IN A WATERSHED**

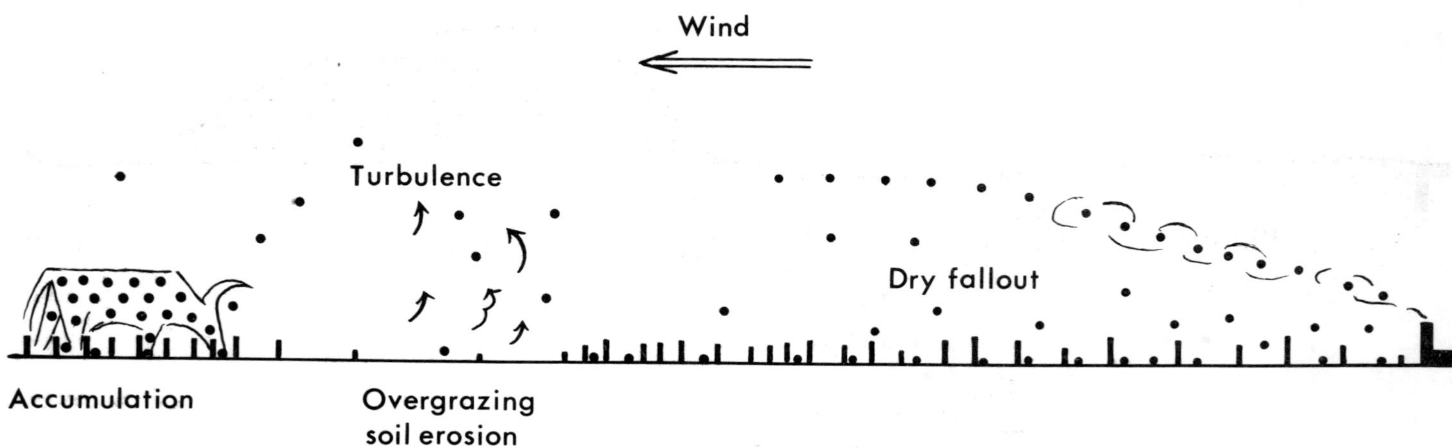

Fig. 8 **EMISSIONS IN AIRSHED**

But are these speculations of any importance to Africa, where there are so few industries? There may be a very real danger in relation to mining activities. In connection with copper mining, both lead and cadmium occur as emissions. In Zaire, Zambia, and the new mines at Selebi Pikwe in Botswana, these problems have to be watched and emissions of heavy metals could be very serious in some range areas. Two effects have to be considered: the productivity of the ecosystem is decreased and the quantity of the meat is affected by residues.

The pollutants and their interactions with dust particles carried by air are also important for man. Lead and cadmium especially may have serious effects on human health, and if these are carried together with dust particles the effect may be synergistic. In an environment rich in airborne particles, the effects of pollutants may be more marked. It is of interest in this connection that health surveys in Botswana have shown very high frequencies of lung diseases. In present mining areas and especially when new mines are opened, health studies should be continuously undertaken as part of the monitoring programme.

Major vegetational regions and the changes in the boundaries between them are of importance in Africa when assessing natural resources or establishing ongoing trends. The desertisation process is largely loss of vegetation cover. The disappearance of tropical forests means a radical change in the height of vegetation cover. These changes could easily be followed by the remote sensing technique using satellites. The ideal instrument for such recordings is the multi-spectral scanner. The results from ERTS 1 (Landsat 1) and the present Landsat 2 are very promising. The climate of Africa, especially in arid and semi-arid areas, allows for regular inventories of the major vegetational changes. It has to be stressed that in the present circumstances this technique offers a very inexpensive method of large-scale monitoring. It is suggested, therefore, that satellite monitoring should be included within GEMS.

Finally, it should be stressed that Africa can fulfil another important function in connection with monitoring, namely, by offering sites for baseline stations. It has been decided that within GEMS a network of global baseline stations should be set up. The function of these stations should be to follow the changes in the global environment. Such stations will measure different types of pollutants and also other parameters which may indicate relevant changes. The stations have to be located in areas little modified by local human activities and they probably also have to be protected from such impact in the future. This means that they have to be in remote areas. At the same time, each station should be located in a place where integration and representation for a large area is expected to occur, as on high isolated mountains. The African continent has many mountains which satisfy the conditions for good baseline stations.

REFERENCES

DESHLER, W.W. 1974 'An examination of the extent of grass fires in the savannas of Africa along the southern side of the Sahara', in: *Proceedings of the Ninth International Symposium on Remote Sensing on Environment*. Environmental Research Institute of Michigan (ERIM), Ann Arbor.

BOWDEN, L.W., HUNING, J.R., HUTCHINSON, C.F. and JOHNSON, C.V. 1974 'Satellite photograph presents first comprehensive view of local wind: the Santa Ana', *Science*, 184 (4141), 1077-8.

BRYSON, R.A. and WENDLAND, W.M. 1970 'Climatic effects of atmospheric pollution', in: S.F. Singer (ed.) *Global Effects of Environmental Pollution*.

KUENEN, H. 1950. *Marine Geology*. New York.

KUKLA, G. and KUKLA, H. 1974 in Environmental Studies 6. 2/3, no.49.

MACLEOD, N.H. 1974. *Dust in the Sahel: Cause of Drought?* mimeo.

PROSPERO, J.M. and CARLSON, T.N. 1972 'Vertical and aerial distribution of Saharan dust over the western equatorial North Atlantic Ocean', *Journal of Geophysical Research*, 77 (27), 5255-65.

RAPP, A. 1974 *A Review of Desertization in Africa - Water, Vegetation and Man*. Secretariat for International Ecology, Sweden (SIES) report no.1. Stockholm.

SCOPE. 1971 *Global Environmental Monitoring*. International Council of Scientific Unions (ICSU). Scientific Committee on the Problems of the Environment (SCOPE) report no. 1. Stockholm.

—— 1973. *Global Environmental Monitoring Systems* (GEMS). ICSU. SCOPE report no. 3. Toronto.

UNESCO. 1974 *Pollution Monitoring and Research in the Framework of the MAB-Programme*. Man and the Biosphere (MAB) report series, no. 20. Paris.

CLIMATIC CHANGE AND ECONOMIC DEVELOPMENT*

JOHN J. HIDORE and MAHDI AMIN EL-TOM
University of Khartoum

Over the span of the Earth's existence, the processes which operate the Earth's environment have reached a state of dynamic equilibrium, and hence so has the Earth's environment. This general statement is applicable to the planet Earth as a whole as well as to the individual regional systems or physical environments. Each regional system seems to have reached a state of internal equilibrium and thus to have developed its own characteristic climate, hydrology, ecology, landforms and economic capabilities. In this sense each region has developed a specific carrying capacity in terms of the human population. The upper limit of the carrying capacity of the region, has always been the most critical aspect as far as human activities are concerned. Hardly any system can be expected to expand beyond the upper limit of its own carrying capacity. No doubt the most critical factor in this respect is the capacity to produce food or, rather, the capacity of the individual ecosystems to produce the green matter which is the basic element in food chains. Of all the possible limiting factors, sunlight, moisture and mineral nutrients may be regarded as the most critical. However, the significance of each of these factors may vary considerably between the individual ecosystems.

Over many parts of the world, the population has tended to expand up to the maximum carrying capacity of the environment and therefore has stayed in equilibrium for fairly long periods of time. The lower Nile valley provides an example of such an area. Here the limiting factor has always been water and thus the population has tended to expand to the limits sustainable by the amount of water available for irrigation and thus has stayed in equilibrium for several thousand years. Such a strict relationship between population growth and environmental capabilities has been disrupted over certain parts of the world so that in some regions the population increase exceeded the maximum carrying capacity of the environment. Under such conditions, varying options have presented themselves. In some cases out-migration was the solution. So the excess population migrated out of the region and this brought about the required equilibrium between the human population and the carrying capacity of the region in question. However, this has not always been the case. In some instances, the pressure was relieved through artificially increasing the carrying capacity of the stressed ecosystem. This was achieved by introducing supplemental nutrients, artificial irrigation, improving seeds or by some other means. Under such conditions the carrying capacity and population increased simultaneously and consequently the ecosystem remained in equilibrium, but at a level somewhat higher than that dictated by the previous physical environment.

*English original.

So over many parts of the world the pressure exerted by the growth of population beyond the upper limits of the individual ecosystem has been relieved either by out-migration or by artificially increasing the productivity of the land, and in both cases the ecosystems concerned have continued to maintain the required equilibrium between the land and human population. Over other parts, however, the population has simply been allowed to increase at the expense of environmental resources. Under such conditions more resources were used than were being made available through natural processes and hence the environment began to deteriorate. This development leads to a reduction in carrying capacity in the face of rising population. The end result has always been great human suffering, especially during periods of adjustment downwards. Such suffering continues until the population is reduced to a level sustainable by the new lower carrying capacity of the environment. In other words, it continues until a new equilibrium is reached.

Such developments suggest correctly that the Earth as a system is in dynamic equilibrium: the Earth system is by no means static but is subject to changes. Some of the changes are a result of external forces operating on the system and some internal. Some of the changes which occur seem to bring about long term effects in the level of equilibrium. These trends are apparently irreversible changes that have been occurring for a long period of time and probably will continue to occur. Other events seem to repeat themselves in cycles of irregular period and varying intensity. For example, the seven lean years followed by seven fat years alluded to in the Bible and Qu'ran is a reference to a moisture cycle. In recent years there has been evidence of cycles with a length averaging 11 years that have been associated with solar flares. These short term fluctuations do not affect the equilibrium level of the environment: they represent minor disturbances and feedback mechanisms operating in the system. These cyclical fluctuations however occur over varying lengths of time. Some oscillations in the environment average hundreds of years in length and others thousands of years. These natural environmental oscillations can result in a more favourable system for the human species and hence increase the carrying capacity, or result in a less favourable system with a reduced carrying capacity.

One of the major controlling elements of any regional system or ecosystem is that of climate. Any change in climate tends to have almost immediate effects on the carrying capacity of the system. For instance, from about 1850 to 1950, a major warming trend took place in the middle and high latitudes accompanied by a large scale increase in precipitation over many parts of the low latitudes, especially over the sensitive semi-arid parts of the tropical regions. These changes greatly increased agricultural production and hence carrying capacity, at a time when the human population was undergoing its greatest expansion up to that time. During this period, the individual ecosystems continued to maintain a natural equilibrium through the intensification of agricultural production but, more critically, through the expansion of cultivable lands into marginal areas. Tropical Africa in particular has witnessed a major expansion of agricultural lands towards the interior of the Sahara desert, while the availability of pasture and water have encouraged the nomads to respond to the increasing carrying capacity of the environment by raising more animals. In this way the tropical parts of Africa have moved towards a higher level of production and the population has increased steadily up to the upper limit of the new carrying capacity of the environment.

However, an environmental change in a less favourable direction should naturally reduce the upper limit of production and therefore reduce the carrying capacity of the ecosystem. This will be brought about through the limitation of the cultivable lands as well as through reductions in crop yields and grazing pasture. Unfortunately such an unfavourable change started to occur in the northern hemisphere around 1950. For the last 20 to 25 years, temperature levels have been dropping continuously over the middle and high latitudes and this has led to a continuous equator-ward displacement of the permanent subtropical high pressure belts. One of the direct and most serious effects of these developments is the evident southward expansion of the Sahara desert and the consequent reduction in the carrying capacity of the drier African environments. Rains here are no longer adequate and thus yields are dropping, cultivable land is shrinking and grazing land is fast disappearing. Now there is a widespread shortage of water and drought has therefore struck much of tropical Africa especially the so called Sahelian zone. Drought constitutes an environmental problem that has plagued mankind throughout historic times, and in all probability long before that.

The term 'drought' always implies a shortage of water supply. However, one may distinguish several kinds of drought, namely: perennial drought, seasonal drought, and intermittent drought. Perennial drought is characteristic of the great deserts of the world. Here water is available only incidentally in the form of occasional rains, or in the form of an oasis or river. The river Nile is an example of the latter case. Seasonal drought characterises the areas where most of the annual precipitation is experienced during a portion of the year leaving the remaining weeks or months rainless. Seasonal drought occurs wherever the climate includes a distinct period of dry weather each year. Intermittent drought occurs whenever the precipitation fails in humid areas, or when, in areas of seasonal drought, the rainy season does not materialise, or is greatly shortened. These intermittent droughts, due to largely unpredictable changes in the temporal pattern of precipitation, are the most serious to mankind. Their seriousness arises from the fact that, unlike the other two types of drought, they disrupt the equilibrium of the ecosystem leading to a marked decline in the carrying capacity of the ecosystem concerned. It is this sort of drought that has caused much of the recent human suffering in tropical Africa. Not only this, but the continuing decline in precipitation is gradually transforming parts of tropical Africa from areas of seasonal drought to areas of permanent drought. Therefore the present inadequate carrying capacities of the African tropical environments are under considerable strain and there is strong evidence to suggest that future environments may have even lower carrying capacities than those of the present day. This is bound to lead to catastrophe in the areas where agricultural development has already expanded to a maximum level during previous more favourable years. The first to be hit are the areas of subsistence production where human population normally expands to, and is strictly controlled by, the maximum ability of the local ecosystem to produce food. So a drop in the production of food should necessarily lead to the reduction of the equivalent proportion of the human as well as the animal population.

It is now quite evident that a climatic shift away from the optimum of 1850-1950 is continuously reducing the natural production levels in tropical Africa by considerably reducing the all-important precipitation levels. This shift started to take place after the local ecosystems had begun to operate at fairly high levels of equilibrium. It is taking place after the local peoples have adapted themselves to extensive agricultural activities and to fairly high crop yields. The negative effects of these unfavourable environmental changes are evident over many parts of Africa. Among the Maasai of Kenya and Tanzania the building up of large herds is considered to be an insurance against losses from drought and disease, the theory being that the more stock a man has, the more he is likely to have left after famine or disease. The folly of building up stock numbers with little or no regard for the declining carrying capacity of the environment was brought home to the Maasai very forcibly during the severe drought of 1960-61 when some 300,000 head of cattle died from starvation in the Maasai district, leaving many herders destitute and dependent on their central Government for survival. That same drought affected many species of wild animals; in fact large numbers of zebra and *kongoni* died as a result of the failure of rains in the 1960-61 seasons.

Another example of these effects concerns the African Sahel zone which has recently witnessed a succession of drier than normal years. Millions of animals, and many people as well, were lost as a result of the Sahelian drought. Of those who survived, a sizeable section is now dependent on international aid while another section moved into other people's land. The continuous migration of people and animals into already crowded areas is creating a complex of social and economic problems and is placing a further strain on deteriorating ecosystems. The western plains of the Sudan provide an example of this situation. The northern part of these plains is occupied by camel-owning nomads while the southern part is occupied by cattle-owning nomads. Traditionally, the nomadic movements have been synchronised in such a way that hardly any contact between these two distinct groups takes place. But recently, the camel owners have found it necessary to graze for longer periods over the southern areas and so they have increased the pressure on the southern grazing lands thus leading to the deterioration or the elimination of pastures that were traditionally reserved for the cattle of the southern nomads. Not only this, but the camel owners used to spend the winter season close to the heart of the Sahara desert in the vicinity of the Sudanese-Libyan borders in order to graze the so-called '*gizzu*' grass but for the last five years the '*gizzu*' pasture has ceased to exist and so the nomads remain further south. This has led to what seems to be an irreversible deterioration of the ecosystem. Also, as a result of competition for pasture and watering places, the frequency of tribal conflicts is increasing very rapidly. The situation is further aggravated by the eastward migration of West African groups especially some Chadian peoples. These are exerting further pressure on the environment. All these developments are manifested in a continuing reduction in the carrying capacity of the ecosystem, a continuing southward expansion of the Sahara desert and a steady shrinkage of the area under dry-land cultivation.

There is little doubt but that drought hazard in Africa is increasing in frequency and geographical extent. The hazard is increasing partly from natural causes but largely from man's activities which tend to aggravate drought conditions by speeding up water runoff and by reducing surface pondage, infiltration, and ground water either through overgrazing or improper cultivation practices. For instance, those areas afflicted by perennial drought, the deserts, are increasing in size. When the soil is dry much higher soil temperatures result. When a gramme of water evaporates it takes with it 540 calories of heat from the soil. Dry soil has no moisture to evaporate and hence no, or less, cooling takes place. Destruction of vegetation on the desert margins exposes the soil to direct solar radiation and wind which dessicate the soil. So this expansion of desert into the Sahel and other areas may not be related to changing atmospheric conditions alone.

The areas which seem to suffer the most from the problem of drought are those which are subject to a seasonal regime of precipitation, particularly where the rains or rainy seasons are short. The shorter the rainy season the more variable it tends to be from year to year. The explanation seems to be related to the circulation system which carries the moisture. Over tropical Africa, the monsoonal winds are perhaps the only rain producing systems. These winds are a result of many complex variables in the Earth's environment and minor changes in the global circulation can shorten or lengthen the period of the flow of air off the ocean. On the other hand, a greater change in world circulation can produce a series of years with shorter than average rainy seasons. It may be worth mentioning here that wind movements over tropical Africa may be affected by developments over areas as far away as the Arctic and Antarctic ice caps. Marked changes in the extent and intensity of the winds may be triggered by relatively small fluctuations in these ice caps. Paradoxically enough, the ice sheets themselves are as sensitive to fluctuations as they are influential in shaping the global climate. It is believed that this is due to the short lifetime and the thinness of the individual ice floes as well as to the extreme seasonal variations in intake of solar radiation over these high latitudes.

So climates are presently changing unfavourably and therefore the capability of the Earth's system to produce green matter is diminishing. Consequently, the demand for food has greatly outpaced the rate of production. This alarming situation is causing worldwide concern, so that in the 1970s the cry for rapid economic development is heard throughout Africa, Asia, and Latin America. One of these demands is for the expansion of agricultural production. Proposals for increasing production per unit of land and bringing more land under cultivation exist in almost every country. At the same time as the cry for rapid economic growth is heard, we hear more and more frequent rumblings of a worldwide shortage of agricultural products. Even though production increases, the shortages on a worldwide basis are increasing. The desire both to expand the economy and the existing level of surplus food supplies dictates that there will be great pressure placed on the Earth's arable land to yield more food and fibre in years to come.

The natural tendency will be to expand agriculture to the limits of the environment that exist at the present, or during optimum conditions. However, because of the natural climatic changes which occur, such practices are fraught with hazards and will inevitably cause great hardships for reasons discussed earlier. It seems vital that future development of regional environments must take into consideration the natural fluctuations which occur in the environment. The limits to agricultural expansion must be established at a level sustainable during periods of climatic downturn, whether in terms of moisture or temperature, depending on which of these factors is the most critical. There are two primary reasons why this must be done. The first is for obvious humanitarian reasons. Whereas the human species has found it acceptable to allow wild animal populations to be controlled by starvation during drought, it is not an attribute of the human species to allow the same thing to happen to his own kind. In the past and, to a limited extent, at present, when starvation faced groups of people it was possible for them to migrate to better areas, or food supplies from other areas were brought to the stricken region. These two alternatives are all but gone. In the first instance virtually all the most productive land in the world is already in use, and there is no new land to move to. In the second, while there have been massive food shipments to distressed areas in the decades since World War II, these are not likely to continue. These massive shipments have been possible due to the huge surpluses of grain that were produced. The world demand for food has increased to the point where the world's supply of surplus grain is of only a few months duration. Large surpluses simply will not be available to move to distressed areas. In the latter part of 1974 two conferences relating to world food problems were convened: the first to consider the total food problem and the second to discuss the possibility of obtaining large quantities of grain to ship to South East Asia and to locate transport to move it.

The second reason why the intensity of economic development must be held to equilibrium level is to prevent further deterioration of the environment. Severe overgrazing or improper agricultural practices during times of environmental stress can disturb the ecosystem to the point where it may not be able to recover following the disturbance. It is this very process of over-development, and destruction of environmental equilibrium which has led to extensive abandonment of the grasslands of the world and to the accelerated expansion of the deserts alluded to earlier.

The only long-range solution to development is to keep land use at an intensity level substantially below that which has traditionally been allowed. Development levels have to be kept close to the lower limits of the carrying capacity of the regional ecosystems. This will safeguard against unnecessary human suffering. But the determination of these critical levels requires a thorough understanding of the physical environment as well as a close monitoring of changing environments, especially sensitive tropical environments. Through such studies it may be possible to throw some light on the nature of the cyclic and non-cyclic changes in tropical climates and it is only through such information that we may be able to determine the real long-range carrying capacities of our regional ecosystems and hence the carrying capacity of the whole planet Earth.

AFRICAN DROUGHT - THE STATE OF THE GAME*

Notes on the Relations between Underdevelopment, Decision-Making, and the Dangers of Reconstruction.

P. O'KEEFE
University of Bradford
and
B. WISNER
Institute of Development Studies
University of Sussex

Following a series of disastrous years of drought in Africa, this article aims to present a synopsis of the current situation. To attempt incisive generalisation encourages ambiguity and inaccuracy but on the other hand the current plethora of publications begs an overview.

The Physical Reality

Most discussion of the African drought focuses upon the region known as the Sahel. In fact this region is extremely difficult to delimit and it has been suggested that it would be better to consider the drought-affected areas in Africa as being regions with less than 700 millimetres annual rainfall. This formulation of the problem in terms of rainfall data would allow an accurate study of drought as a climatic phenomenon and an assessment of the relationship between the natural hazard drought and the famine experienced in the Sahel (O'Keefe, 1974a).

Such a regionalisation would also allow the inclusion of areas in Ethiopia, Kenya and Tanzania where, although the ecology might differ, the process remains fundamentally the same (Berry et al., 1972). Few would now deny that there has been climatic variation on a global scale (Lamb, 1974). One authority has analysed climatic variations for Africa (Winstanley, 1973a; 1973b) demonstrating that, since 1960, the summer monsoon rainfall of the Sahel zone shows a high negative correlation with the winter-spring rainfall of the Mediterranean. During the 1960s the circumpolar vortex expanded and the meridional circulation in the tropics contracted, giving heavy winter-spring rainfall over the Mediterranean and the Middle East, and low monsoon rainfall in the Sahel zone. The change since 1960 is interpreted as a continuation of trends first apparent in the 1920s. Winstanley argues that the monsoon rainfall is at a low point of an irregular fluctuation superimposed on a longer downward trend. The present drought extending from Mauritania to India is a result of the cumulative effects of a decade of below average rainfall with less and less rain each year. He estimates that, because of the low mean annual rainfall and the strong north-south precipitation gradient of the Sahel, a southward shift of the climatic zones through one degree of latitude over

*English original.

100 years means a decrease in rainfall of 40-50%; furthermore he estimates that there has been a southward shift of the isohyet range by approximately 9 km per annum. On a microcosmic level Bradley examined the implications of the 1973 rains at Samaru, Nigeria, and argued that the low early season rainfall would have a detrimental effect on the annual total and, therefore, on agricultural production (Bradley, 1973). Early season deficit would produce a poor harvest because for most crops good growth is essential during the first month after germination.

It is worth stressing that, because of the shortage of climatic data, no watertight case can be presented to calculate the significance or severity of the drought. While Lamb and Winstanley argue that increasing dryness is the long term prospect for the Sahel, others would suggest the possible existence of long periodic cycles. The important point to stress is that in the Sahel the human response to low, and highly variable, rainfall has been nomadic pastoralism and dry farming - both of which reflect a functionally adapted way of life. That the population has been unable to cope with the present shortage of rainfall indicates the severity of the drought and the lack of alternative available responses.

Parallel to the arguments concerning climatic variation are the myths that surround the environmental processes of *desertification* and *deafforestation*. We would argue that these concepts tend to blur rather than clarify the discussion of environmental degradation. At this juncture it is worth emphasising two points:

(1) That the only area of real downgrading desertification is the intermediate region between the area where the grasses are predominantly annuals and the area where they are perennials (500-700 mm rainfall). There seems little reason to presume a southerly drift of a frontier of sand.

(2) That deafforestation often means the lack of bush savanna but bush savanna has less potential utility in terms of protein generation than comparable grassland (O'Keefe, 1975).

In summarising current knowledge of the environment it appears that three things need to be emphasised. First, it seems impractical to delimit the drought zone according to vegetational phenomena - after all drought is concerned with the amount of rainfall *not* received. *Any area receiving less than 700 mm of rainfall should be regarded as drought-prone*. Secondly, it should be emphasised that although drought has been severe in the past few years, *most authorities would expect that this is only the beginning of long-term drought prospects for Africa;* this has ominous implications for future development. Thirdly, *the area which will suffer most ecological degradation seems to be an area between the 500 and 700 mm isohyets,* an area which supports both pastoral and arable economies but which is often considered to be beyond the scope of relief activities.

A Brief Description of the Drought-Famine Process

It is self-evident that the form of peasant resource management in dry areas is functionally adapted to the environment. Any functional adaptation requires ongoing change as the range of the experience-response system expands; change implies the necessary political power to enforce further adaptation and it is the absence of executive power that has made peasant farmers so vulnerable during the current drought. If one assumes that these peasant landuse systems were once successful and efficient, then one needs to reconsider their recent history to understand why they should be so vulnerable in the current crisis.

Adaptation can be viewed within the framework of game theory. It is assumed that the limitations of environmental potential and social organisation allow for an optimum *strategy* to be followed at a specific time. Strategy, however, is quite distinct from the *players* and both strategy and players are distinct from the *rules* within which strategy is evolved by generations of players. If one reviews the current crisis against this theoretical framework, the following points come to light.

a) In 27 of the 39 years, from 1920 to 1958, rainfall in the semi-arid areas of Africa was above the long period average; this change would necessarily encourage the evolution of a different environmental strategy by the peasant farmer.

b) No alternative environmental strategy was evolved because of the impress of colonialism. Colonialism had far-reaching effects but these may be summarised in game theory terms as follows:

   i. The composition of the *players* changed, possibly as death and disease control were introduced without concomitant birth control, certainly insofar as colonialism accelerated the process of, and brought new forms to, rural stratification - thus creating classes of antagonistic players: landless, small, middle, rich peasant, etc. This has sharpened the contradiction between some strata of the peasantry as a result, e.g. pressure to increase cash crop acreage forced sedentary peasant farmers to encroach on grazing land previously used by pastoralists. Strictly speaking the 'game' became less a game of man-against-nature and more a game of man-against-man-against-nature. Furthermore, land pressure (e.g. in the Kenyan Highlands) has forced increasing numbers of very poor peasants down into the semi-arid plains - thus providing yet more 'new' players and exacerbating competition among players in the dry zone.

   ii. The *strategy*, previously a functional adaptation, was rendered inappropriate by colonial legislation which in some parts emphasised statutory and individual land tenure against traditional patterns. Alternative pastures and virgin land were reduced; the land frontier began to close.

   iii. The *rules* changed as the economy was assimilated into an era of market force manipulation. Consequently, the resource base itself was eroded. Deterioration of the resource base was the direct and indirect result of syphoning off surplus value from the rural areas - a process which spans both colonial and neo-colonial periods (Suret-Canale, 1971; Amin 1973; Brett, 1974; Meillassoux, 1974; Leys, 1975; O'Keefe & Wisner, 1975).

Despite changes in the composition of players and rules, to a significant degree the *strategies* of landuse, herd management, etc. employed by peasants in arid regions - especially those employed by the poorest peasants in these areas - have remained unchanged. In order to understand why this is so one must analyse the way in which colonial and neo-colonial modes of production have added constraints to the objective and subjective frameworks within which the peasant makes decisions. This analysis requires, in turn, a closer look at the application of game theory to small-scale agriculture.

Gaming models have been used for some time by students of agricultural decision-making in cases where it is assumed that 'payoffs' from a given set of agricultural decisions are not simply given by some input-output function, but are also affected by a given 'state of nature' (Gould, 1963; 1965). Various assumptions are made about possible states of nature, their frequency distributions, and the state of the farmer's knowledge of these distributions. It is further assumed that the farmer's response to a given state of nature is chosen by him from an array of possible actions of which he is aware through personal or cultural experience. Changes in the farmer's objective conditions throughout the colonial and neo-colonial periods - especially the rise of the capitalist mode of production - have worked to constrain this array of possible actions or range of choices. Constraints have been applied in the following ways:

i. Direct capital penetration into the more arid parts of many African countries meant direct expropriation of the land resource and an absolute decrease in the land base of the livelihood systems of poor peasants in these areas. For example, the introduction of coffee to the Mt. Kenya highlands in the 1950s created a 'coffee-bourgeoisie' in the highlands, who invested widely in cotton farms on the drier lower slopes of the mountain in the 1960s and found additional investment outlets even further out into the dry plains when they obtained control of more than 800 km^2 of land which was then 'enclosed' and named Meru Game Park. The effect of highland capital's penetration via investment in cotton and the game park was to cut by at least one-third the land available to poor farmers in the lowlands who had combined dry-farming with transhumance herding in what had previously been 'common' land.. Such examples abound in Africa, and such decrease in the livelihood base of poor peasants *usually* results from the establishment of 'modern' ranching by agri-business and large-scale irrigation schemes.

ii. Where direct capital penetration has not yet taken place the orientation of the national economy towards the needs of the capitalist mode of production generally means that the poor farmers of dry areas have severely limited access to new strategies which could enlarge their range of choice. It is well known that such inputs as credit, extension, marketing - not to mention infrastructure - are maldistributed in Africa according to colonial patterns favouring so-called 'high potential' areas. These patterns have generally been perpetuated by the independent governments.

iii. Where the programmes of independent African governments have attempted to enlarge the peasant's range of choice, these programmes have never tried to deal with the whole problem of regional underdevelopment but have focused on a small number of discrete superimpositions on the livelihood systems of the neglected area, e.g. a new seed variety, some boreholes, a road. These partial solutions, precisely because they are partial, often conflict with the pattern of survival behaviour evolved in the neglected region. The net effect is a further decrease in the poor peasant's range of choice.

The subjective framework of peasant decision-making has been attacked by lack of respect for traditional science and technology. This lack of respect - often bordering on denigration - is found throughout the civil service, educational systems and national bureaucracy. It rejects 'people's science' wherever it appears, be it traditional flood-retreat agriculture folk medicine, or house-building. These attitudes - born in the colonial period, but perpetuated by continuing transfer of technology from the world's metropolitan centres and by an educational system designed to produce manpower for the capitalist sector - are major obstacles to the emergence of *hybrid* technology which would draw from both traditional and modern sources, thus enlarging the peasant's range of choice.

The combined result of these various sorts of constraint on peasant decision-making is at best stagnation of the peasant range of choice - failure to enlarge the area of possible action which is a *sine qua non* of any new peasant strategy. At its worst it has meant an actual shrinking of the range of peasant choice, often brought about by the atrophy of various cultural adaptations based on kinship, such as mutual aid in land clearance and harvesting, or cattle trusteeship. As constraints on strategy build up it becomes more likely that, at some point, stress which previously might have been dealt with becomes, in the new situation, too much for the peasant livelihood system to bear, as has happened in the Sahel and Ethiopia. Even where catastrophe is not the result, increased vulnerability to stress often produces a highly simplified - one might even say 'pathological' - type of peasant decision-making. The phenomenon of *decision-pathology* is, in fact, a greater challenge to long-run development planning than the overt economic dislocation seen in the Sahel. The examples discussed below demonstrate the magnitude of this challenge.

Failure is common in all agricultural activity whether one judges by the rates of bankruptcy in the USA or the frequency of hunger in Africa. Thus no matter what one 'plays' against the opponent, nature, there is a good chance that one will 'lose' something. Even granting that the gaming approach helps one anticipate the order of magnitude of probable losses (useful if one is a grain marketing board, questionable if one is a small farmer), it is clear that the main focus of interest naturally moves on beyond the immediate payoff. Given such and such a payoff, what does the farmer do? He still has to keep his family fed and otherwise provided for. Therefore the *critical* questions lie between 'rounds' in the game with nature. By joining a series of gaming decisions, one produces a rudimentary 'learning model'. However, each 'round' involves the same *type of decision*: the farmer makes some judgement about the probability of a state of nature; chooses inputs according to some algorithm which avoids risk and seeks utility to some degree or another; and receives an output, the 'payoff', which will somehow modify both the farmer's knowledge of probable states of nature and the choice of algorithm. One can hypothesize a period 'in-between rounds' when a decision of another type is taken. This second type of decision simplifies the situation, seeks certainty by consideration of a small number of 'sure bets' which cultural adaptation has pre-selected to be the most likely to provide a livelihood once 'losses' have exceeded some culturally and environmentally specific threshold.

The closer to a positive quantity the payoff is, the more likely that the next decision will be sensitive to the next state of nature. The closer to the threshold loss the payoff is, the more likely that the next decision will not be sensitive to the next state of nature. Herein lies a partial explanation of the fact that poor peasant strategy has often remained constant while players and rules have changed. Progressive deterioration of the peasant resource base, decrease of reserves, increasing constraints - everything, in short, which contributes to increased vulnerability to environmental stress - cause the peasant to rely more frequently on such over-simplified decisions. The additional irony is that once colonial and neo-colonial patterns force the peasant into such modes of decision-making, social scientists in the service of colonial or neo-colonial interests *lay the blame* on the peasant himself for being 'irrational'.

Dodoma (Tanzania) April, 1969: rain has been poor, grain supplies are quite low, some threshold has been passed and the threat to the family is so great that an unprecedented sale of cows has taken place which peaks in December with 3000 cows sold in Dodoma district. Since 1966 the monthly sale of cows had fluctuated around 600. The 1969 sale was the result of thousands of very serious decisions on the part of peasants. To sell one's cows is to cut into core reserves. Indeed an all time low in total cattle sales in mid-1970 would suggest that the breeding nucleus of some herds was affected. This liquidation of reserves represents a different type of decision from the decision to shift one's cultivation into an available lowland swamp , to plant extra cassava, or to pray. These latter are adjustments to soil moisture extremes and associated losses within the range bounded by some normal positive payoffs and the threshold negative payoff.

The decision to migrate and the decision to sell cows as opposed to other livestock are both generally decisions which are less sensitive to the series of successive states of nature. The members of this set of emergency adjustments are few. They are resorted to only when the level of reserves of a farm family (including livestock, grain,availability of credit and gifts from more fortunate kin, savings, and capital goods) is very low with respect to relatively fixed minimum family needs and relatively high possible additional losses. One has witnessed in the Sahel the extremes to which the necessary combination of increased vulnerability and increased pathology of decision-making can lead. What are the dimensions of that extremity?

One hardly needs to emphasise the lack and inaccuracy of demographic data in this particular situation. It is sufficient to state that if one accepts a catastrophe as a situation in which more than 100 people die and/or 1,000 are injured and/or $1,000,000 worth of damage is sustained,then this disaster is parallel to recent disasters in Bangladesh in terms of magnitude of death, disability and destruction (UN Protein Advisory Group, 1974). Six years of drought have brought the sub-Saharan areas of the Sahel, parts of Ethiopia, and Somalia to nothing short of catastrophe. More than 25 million people have been affected and initial estimates for West Africa suggest that more than 100,000 people have died (Sheets and Morris, 1974). Although these figures have been queried this seems the most realistic estimate.

Estimates of cattle losses are also high; it is thought that Mauretania has lost 1.6 million cattle from a total of 2.1 million in the last two years.

Cattle losses are placed around 50% in Senegal, 50-80% in Mali, 50-100% in Upper Volta, 80% in Niger and 90% in Chad. FAO estimates the 1973 cattle loss at 3.5 billion head valued at $400 million. This suggests (according to Oxfam, February 1974) a loss of 25% in the Sahel herds. Crop losses have varied between 50-70% (± 20%).

General statistics tend to be misleading,so other data sources are necessary. The best indications of the population status can be gathered from the nutritional surveys completed during the last two years (Seaman et al., 1973). Several pessimistic conclusions emerge from these surveys. First, that although the anthropometric and clinical evidence presented indicates that the population is on a low nutritional plane, the general situation is not unusual for rural Africans in similar environments. Secondly, the poor harvests of 1972, 1973 and 1974 have produced a serious economic situation in which nomad pastoralists have been more severely affected than sedentary farmers, although the latter have also suffered; long term drought prospects would prove disastrous. Thirdly, it is not possible to state whether or not the people who have received free distributions of cereals have actually depended upon them. Fourthly, it is accepted that death seldom results from malnutrition *per se* but because a malnourished population is susceptible to endemic and epidemic diseases. In this context it is worth noting a major increase in deaths from measles in West Africa. Fifthly, and most importantly, the continued erosion of food stores stock-piled against the possibility of long-term drought, indicates that a more permanent solution than relief aid must be found. With current estimates of malnutrition suggesting that 10% of the sedentary and 50% of the nomadic population is undernourished there is an immediate need for long-term planning.

Such a stress situation encourages wholesale migration and to some extent increases the difficulties of relief distribution and rehabilitation. Traditionally two migratory patterns were discernible:

i. A rural-urban movement for wage labour.
ii. A seasonal movement of semi-nomadic population utilising traditional routes in search of water and pasturage for their animals.

None of the traditional alternatives, which are spatially finite in a subsistence community experiencing stress,are available. It is useless to increase the labour input, useless to search for fertile unoccupied land, difficult to establish external linkages to provide secondary sources of income, pointless to emigrate to equally barren areas and futile to implement population control programmes which do not have immediate benefit (O'Keefe, 1974a). As a result of the stress situation and the absence of alternative strategies major migratory variations have occurred. Namely:

i. Migrants have deviated from traditional itineraries: nomads are arriving in urban areas, and some have been arriving in areas further south than normal, thus causing political friction.

ii. Increased numbers of people are migrating. Entire families have been seen on routes usually followed only by male herders and their animals.

iii. Reasons for migrating have changed. No longer are migrants searching for pasturage but rather for government-distributed food.

iv. The normal seasonal migration is not being balanced by the usually predictable return. Many nomads will remain within the urban area (Sheets and Morris, 1974).

Such a situation indicates that the drought has turned to famine even if, at the moment, it is impossible to quantify the exact extent of social and economic dislocation.

Reaction to the drought and growing famine was extremely slow. It took five years for the plight of the people in the Sahara to become newsworthy - so newsworthy that three colour magazines in UK have devoted space to the topic. Many articles have emphasised that modern techniques have been insufficiently mobilised to give warning of the approaching famine. The FAO/WFP (Food and Agriculture Organisation/World Food Programme) Early Warning System gave the first intimation of critical crop failure and exceptionally poor harvests only in September 1972; it was not until the end of February 1973, after the Economic Commission for Africa Conference of Ministers that an FAO/WFP Inter-Departmental Working Group on the Problems of the Sahelian Zone was established. If the world press was tardy in appreciating and reporting on the drought, the response of the international organisations was equally slow even though they had the necessary structure to ameliorate the effects of drought.

What is more worrying than the slowness with which *relief* was made available is the conservatism of present plans for *rehabilitation* and *reconstruction* in the area. Since aid for rehabilitation and reconstruction is only a special case of aid in general, there is always the danger that such programmes will simply reconstruct the *status quo ante* (Baird et al., 1975). Large scale river basin development and agri-business ranching clearly contain that danger, especially when the logic behind them follows some variation of Lonrho's questionable formula: Arab capital plus European technology plus African labour (see Comité Information Sahel, 1974, 157-170). The reconstruction of the *status quo* can only mean further increase in the vulnerability of the poorest strata of the peasantry in Africa's dry areas. The alternative is programmatic rehabilitation and reconstruction of the area built on releasing the productive energies of the inhabitants of the dry zone. Such programmes demand new modes of production, and they in turn demand new social relations. This is an essentially political activity of the kind that has been underway for some time in northern Mozambique (Wisner, 1975). In addition such programmes would be built on clear recognition of the following facts:

i. The problem of drought is basically a problem of poverty, of continuing regional economic maldistribution and underdevelopment *within* countries, leading to increasing vulnerability of the poor in these regions.

ii. Any large scale reconstruction initiative (including agri-business investment) carries with it the danger of displacing or, at the minimum, proletarianising the poor peasant (both sedentary and nomad) into increased vulnerability to drought.

iii. Peasant perception and utilisation of local resources can be (and usually is in the absence of the constraints listed earlier) very sensitive to states of nature. Hybridisation of folk and modern technology would thus seem a strong foundation for development in the dry zone. However, new organisations of production are called for, especially small-scale cooperatives amongst peasant producers.

REFERENCES

AMIN, S. 1973 *Neo-Colonialism in West Africa.*

BAIRD, A. et al. 1975 *Toward an Explanation of Disaster Proneness,* Bradford University, Disaster Research Unit Occasional Paper no. 10.

BERRY, L. et al. 1972 *Human Adjustment to Agricultural Drought in Tanzania: Pilot Investigations.* Clark University, Natural Hazards Research Paper 19.

BRADLEY, P. 1973 'The delayed start of the 1973 rains at Samaru, Nigeria', *Savanna,* 2, 78-81.

BRETT, E. 1974 *Colonialism and Underdevelopment in East Africa.*

COMITE INFORMATION SAHEL. 1974 *Qui se Nourrit de la Famine en Afrique?* Maspero, Paris.

GOULD, P. 1963 'Man against his environment: a game theoretic approach', *Annals of the Association of American Geographers,* 53, 290-7. *also in* A. Vayda (ed.) *Environment and Cultural Behaviour,* 1969.

—— 1965 'Wheat farming on Kilimanjaro', in *General Systems Year Book.*

LAMB, H.H. 1974 'Last year's weather', *Geographical Magazine,* 46, 445.

LEYS, C. 1975 *Underdevelopment in Kenya.*

MEILLASSOUX, C. 1974 'Development or Exploitation: is the Sahel Famine Good Business?', *Review of African Political Economy,* 1, 27-33.

O'KEEFE, P. 1974a *The Ecological Background to the Sahel Drought.* Bradford University Disaster Research Unit, Series II, no.1.

—— 1974b *Gakarara - a Study in the Development of Underdevelopment.* Ph.D. thesis, University of London.

—— 1975 *African Drought - a Review.* Bradford University Disaster Research Unit Occasional Paper no.8.

—— & WISNER, B. 1975 'Underdevelopment in East Africa: a Kenyan Case Study,' *African Environment/Environnement Africain,* I.3. forthcoming.

SEAMAN, J. et al. 1973 'An inquiry into the drought situation in Upper Volta', *The Lancet,* 2, 774-8.

SHEETS, H. & MORRIS, R. 1974 *Disaster in the Desert.* Carnegie Endowment for International Peace, Washington.

SURET-CANALE, J. 1971 *French Colonialism in Tropical Africa 1900-1945.*

UN PROTEIN ADVISORY GROUP. 1974 *Recommendations. August 1974.* Mimeo. United Nations, Geneva.

WINSTANLEY, D. 1973a *Nature,* 243, 464.

—— 1973b *Nature,* 245, 190.

WISNER, B. 1975 'Famine Relief and People's War in Africa,' *Review of African Political Economy,* 3, forthcoming.

WATER RESOURCE PROBLEMS IN AFRICA*

LEKAN OYEBANDE
University of Lagos

Introduction

Just as life cannot exist without water, so also our socio-economic system cannot survive or attain hoped-for future excellence without water. Water is a prerequisite of development in many sectors of our economy including industry, agriculture - especially for irrigation in areas of limited or seasonal precipitation - and electric power generation. Water stored in rivers and lakes can provide a basis for transportation and communications. It is indispensable in domestic activities, and if carefully planned and managed can also help meet the growing need for outdoor recreation. The importance of water has been brought home to us in Africa by the recent drought crisis, which has affected many parts of the continent with catastrophic results, including temporarily halting the progress of some national or regional development plans (Oyebande, 1973). The development of this resource is therefore one of the most important and critical problems of contemporary Africa and will become increasingly so as the population increases year by year. Hand in hand with this goes the development of other environmental resources associated with water management - the lands, the forest, the fish and wildlife - all comprising a part of the natural wealth of the African environment.

The developing economies of Africa demand the full exploitation of their natural resources in order to provide a sound economic base for balanced growth in other sectors. Since water plays such an important role in the life of a nation, demand for the resource has increased tremendously with increasing population and urbanisation and it is fast becoming an item in short supply. As man's need for food and fibre grows, supplemental irrigation becomes increasingly important in order to make the maximum use of the environment. Water is in fact a major constraint on the number of crops that can be grown in a year. Even in countries where there is abundant rainfall, the false impression of an unlimited supply of water is fast disappearing for, as development of the various sectors of the economy picks up momentum, demand for water increases very fast, resulting in acute local shortages. The water authorities thus discover that they have to go further afield to obtain adequate supplies.

In their struggle to develop their national economies African governments must utilise their water resources efficiently by scientific planning based upon systematic studies in the field of hydrology. Whilst this goal is yet to be realized in most cases, it is fortunate that many are becoming aware of the significance of providing adequate water supplies not only for domestic, industrial, commercial and other needs in urban areas with their large concentration of people, but also for improving rural welfare in order to reduce the social inequalities between living conditions in the two types of environment.

In Tanzania, for example, some 8.7% of the total anticipated central government expenditure (Warner, 1969) was allocated for urban and rural water supply during the first 5-year development plan period (1964-69). During the second development plan (1970-74) the nation set itself the task of ensuring that its entire rural population was provided with an adequate water supply within a period of 40 years (Warner, 1970). This would cost the government about 8.9% of the total development investment for the period. Similar investment in water development was enshrined in the Kenyan government's development plan for 1970-74. The sum of K£ 1.135 million was allocated for water supplies. This represented some 4% of the total planned expenditure and ranked ninth out of a total of 37 development programmes (Klasse-Bos,1969). In 1964 only one-sixth of the population of Ghana enjoyed good drinking water, made available through 74 public supply systems. During the second development plan (1964-70) the nation envisaged an expenditure of G£ 24.4 million on water and sewerage (Ghana Planning Commission, 1964). This amount represented 5.1% of the total planned expenditure. The sum was to be spent to provide urban supplies through Akosombo and the Kpong extension to increase the supply to Accra and Tema to about 22 million gallons per day (m.g.d.) (c. 100,000 m^3 per day). Other towns as well as rural areas were also catered for.

Similar efforts to develop water resources have been observed in many other African countries, and Nigeria is no exception. During the 1962-68 development plan period, the total capital actually spent on water supply and sewerage was £24.7 million or 4.4% of total actual capital expenditure (Nigeria Federal Ministry of Information, 1970). The corresponding values for the 1970-74 plan period were ₦51.7 million and 5% respectively. Water for all purposes was to be supplied to various towns in all twelve states and to the Federal capital of Lagos. For instance, by 1969 (phase I) water supply in Lagos had increased to 45 m.g.d. (205,000 m^3) at a cost of ₦5.8 million. This is to be further increased to 100 m.g.d. (450,000 m^3) during the third national development plan, to serve a population of about 2.1 million.

Although much effort and money has been invested in water resource development, the net results achieved have not been easy to evaluate because the objectives of such development projects are often not clearly and explicitly stated. Furthermore, the efforts by governments to provide adequate supplies can be, and have been, seriously undermined by the general attitude of the people that water is nature's free gift which the authorities should provide in unlimited quantities. Thus in addition to their difficult task of investing in water resource development to furnish adequate supplies, the authorities must also educate the people and improve their perception of the water resource problem of the various parts of the continent.

Water Resource Problems: a Typology

In Africa, as in most other environments, the basic water resource problems are of three main types, namely: shortage, flooding and pollution. Water shortage, which may be brought about by meteorological conditions or by growing demand for water (or a combination thereof), remains a serious problem in both urban and rural areas of Africa. On the other hand flooding and pollution hazards are most seriously felt in urban centres where large concentrations of people and economic activities both cause and aggravate the effects of these problems.

But perhaps the most important and destructive water resource problem of the African environment is the control and effective utilisation of surplus seasonal waters, both for areas where they originally fell as rain and for other needy areas. This involves the problems of the comprehensive planning and operation of water development projects, and adoption of conservation methods aimed at reducing avoidable losses and wasteful usage, as well as the difficult problem of regional cooperation, when necessary, among the politically fragmented countries of the continent. Several African rivers and lakes flow or spread across two or more countries while ground water flow is also often an international phenomenon. This need for regional cooperation for the effective development of Africa's water resources, combined with lack of basic hydrological and hydrometeorological data, further aggravates the already intricate water resource problems of the continent.

Water Shortage

Precipitation is the source of the world's supply of fresh water. As it falls a portion of it immediately becomes part of existing stream flow, while evaporation and transpiration reduce the amount of water obtainable from precipitation through surface and ground water sources. Climate is thus an essential factor in man's management of available water. To understand the problem of water shortage in Africa it is necessary to bear in mind the pattern of distribution (spatial and temporal) and the intensity of rainfall over the continent. The problem of water shortage in relation to rainfall patterns and other factors is of varying severity over the African continent, but in general the difficulties of tropical agriculture are undoubtedly due in part to water problems, of which insufficient precipitation is one.

The distribution of precipitation is geographically highly uneven. Over 36% of Africa's entire area there is too much rainfall, 16% is arid, while the annual distribution of precipitation can only be considered normal over 22% of the continent (Szentes, 1968,p.75). Many areas of Africa that are considered short of water receive rainfall in sufficient quantity but suffer because of its uneven distribution in time. The water problem does not arise so much, therefore, from deficiency of total rainfall, which over large areas of the continent averages 750-1200 mm p.a., as from its seasonal character and its variability from year to year. It is for this reason that annual rainfall figures are misleading. Other decisive factors affecting water supply are the constantly high air temperatures and the ability of the ground to absorb water. Thus in the Republic of South Africa nearly half of the total precipitation flows through arid regions. Lack of suitable storage sites and irrigable land limits the extent to which this flow can be used to supplement scanty rainfall (de Gregori, 1969, pp. 56-7). Similarly in Namibia the Kalahari has been termed 'desert' because of the absence of surface water resulting more from its sandy soil than from an absence of rainfall.

Major African rivers such as the Nile, Tana, Zambezi, Niger, Volta, and Senegal do carry large volumes of water through more or less arid zones, but their flow is rather irregular. Thus, harnessing water resources for irrigation is an essential factor in African agricultural development both for the regions of Mediterranean climate and for the arid and semi-arid parts of the tropical zones. This perhaps accounts for the emphasis placed on river valley development for agricultural and industrial purposes by countries such as Ghana, Nigeria, Sudan, Egypt, Zimbabwe and South Africa. Attempts to harness the flow of African rivers have varied in scale from simple flood control schemes (Niger valley types), diversion dams with water distribution schemes (Gezira type), to irrigation networks based on multi-purpose storage dams (Maghreb, Egyptian types), the latter being the most expensive of all.

Until very recently, virtually all efforts by African governments in the field of water resource management were directed toward relieving shortage in urban centres. In spite of this concentration of effort, many African towns still have no reticulated water so that people have to rely on sources such as rivers, wells and boreholes for their supplies. Even in towns and cities where large sums of money are continually being spent to provide more water, frequent shortages still occur - for example, in Dar es Salaam, Nairobi, Lagos - due to ever-increasing demand for water by a rapidly expanding population and ever increasing economic activities. Consequently, water supply for domestic use still falls short of the desirable minimum, not to mention water for industrial and other uses. The normal daily water requirement for an adult for alimentation only is 2.5 litres (4 pints). Under hot tropical conditions this amount may increase fourfold or more. In addition much greater quantities are required for domestic activities. Altogether the domestic water needed in homes with modern inside sanitation is at least 115 litres (25 gallons) per head per day (Isaac, 1965). After allowing a conservative proportion of 40% for water needed by industry, public institutions and parks and for watering household gardens, per capita consumption of about 68 litres (15 gallons) and 82 litres (18 gallons) per day respectively were obtained for Jos town and the city of Lagos, both in Nigeria (Oyebande, 1974b; 1974d). These urban centres are amongst the group of African towns most abundantly supplied with reticulated water, yet the available supply is much below the recommended minimum of 115 litres per head per day.

Furthermore in these and other African towns, many industrial establishments have had to develop their own water supply sources to supplement what they can obtain from public supplies. Such auxiliary sources often come from boreholes. In the continual struggle to obtain adequate water for domestic and industrial uses in African urban centres, other uses of water - water for recreational purposes for example - have not been given much consideration, yet these aspects can be important in urban areas for both aesthetic and psychological reasons.

In almost every African country, many have cried out against the long standing social injustice of spending large sums of the taxpayers' money to cater for the small proportion of the population living in towns and cities to the neglect of the masses who live in rural areas without clean potable or even dependable water supplies. Often, however, it is argued that the urban bias is justified on the grounds that the concentration of large numbers of people in cities constitutes a great potential health hazard should a water-borne epidemic disease break out. The fact remains, however, that water shortage in the rural areas causes real hardship and may involve wastage of many man-days spent in looking for water and carrying it over great distances. These frequent shortages, especially during the dry season and during prolonged droughts, cause the price of water to fluctuate very greatly in rural areas. There are parts of East Africa - in Tanzania for example - where due to acute shortages of water the price is as high as one shilling per *debe* (4-gallon tin) i.e. over 40 times the price of water in Dar es Salaam. In that country (Klasse-Bos, 1969), about 90%, i.e. 10 million, of the rural population were considered to have inadequate water supply in 1969.

As mentioned earlier the need for the improvement of rural living conditions to diminish the gap between town and country is being increasingly appreciated, thanks in part to the recent severe drought disaster which affected large areas of the Sahelian zone of Africa. Only a few countries, however, have been able to mount cohesive programmes of action in this respect. Nevertheless, the construction of rural water supply systems varying from simple wells to small storage schemes with elementary treatment is being undertaken in a number of countries. The last seven years or so have also witnessed a much greater priority being given to rural areas, although such priority varies from country to country. Mention has been made of Tanzania which has initiated an ambitious rural water supply programme aiming at supplying good quality water to all its citizens by the end of a 40 year period. Ghana has embarked on a programme of rural water supply through new rural community schemes and extension of existing water supplies. The schemes take the form of providing certain areas with boreholes and wells where ground water is available, especially in the Southern areas; in Northern Ghana and other dry parts such as Brong-Ahafo and the Lower Volta, the basis for development is the provision of reservoirs and artificial ponds. Kenya in 1969 placed a committee in charge of its rural water supply programme to formulate a consistent policy for an enlarged and improved scheme. The provision of rural water supplies was to be phased over a 30 year period. During the first twenty years all parts of the country would be provided with basic communal supplies, while in the final ten these schemes would be upgraded to provide individual house connections.

Lack of rivers is one of the main features of the central part of North Africa, that is to say of Algeria, Tunisia and Libya. Water courses are no more than periodic torrents and wadis, except for the Chalif in Algeria and the Madjerda in Tunisia. Nevertheless in Libya very considerable sums are being spent to provide water supplies in particular areas of the desert. The Kufra oasis project is already in operation on a pilot scale (ECA, 1971). Sheep are now being reared on lucerne grown by virtue of the large quantities of water now made available in an area which was predominantly arid and hitherto very poorly developed. Finally, in Nigeria efforts and money have been spent during the periods 1962-68 and 1970-74 to improve and expand rural water supplies as well as,through catchment-wide hydrological studies and the setting up of water corporations in most states,to oversee water supplies in most areas. These water corporations receive a substantial part of their investable capital from the Federal government.

Flooding

Floods may be defined,firstly,as exceptionally high rates of discharge in water courses and,secondly, as inundations of normally dry land (Wolf, 1966). It is also convenient to differentiate between floods in natural and artificial catchments. The latter,comprising urban areas, airfields, and other large systems where man has covered a substantial proportion of the catchment by paving, roofs etc., which shed rain water directly into impervious drainage systems,are often extensively and severely affected by floods. In Africa flood hazards are greatest in urban centres located in the coastal plains where low, elevation and gentle slopes combine with intense coastal storms and tidal upsurges to cause especially damaging inundations. This is not to say that flooding is not a real hazard in the country districts of the continent. Only recently (February 1975) the flow of the river Limpopo rose to dangerous levels in Mozambique and flooded large areas of the southern region of the country washing away roads, railways and other developments while hundreds of hectares of farmland were endangered by the damaging flood. Large scale rescue operations were needed to salvage lives and property trapped by the furious flood waves. The irony of water resource problems in the African environment is that frequently while water is still in short supply, many towns and their environs are subject to flooding of increasing frequency,magnitude and extent. Such floods have in many cases caused damage to roads, buildings and other property as well as totally disrupting transport, marketing and normal social activities for days at a time. Flooding has on occasions constituted a serious threat to human health and lives. Thus the flooding problem, seen in the right perspective, is part of the dual problem of control and effective utilisation of water resources.

Little information is available on the flooding problem as it appears that hydrologists and the authorities are still preoccupied with the problem of providing adequate water supplies. The results and recommendations of a preliminary report on a study of the flooding problem in greater Lagos (Oyebande, 1974e) are applicable,however, to other humid parts of Africa where the flooding menace is a frequent occurrence. In the humid areas of tropical Africa, especially the coastal belts of the equatorial region, rainfall intensities are usually high. An intensity of 100 mm per hour can be expected for one-hour rainfall once in 15 years if not more frequently (Oyebande, 1974c). Occasionally during the rainy season there may be periods of three or more days of more or less continuous rain. The occurrence of such antecedent rains builds up soil moisture content thus causing high residual surface flows prior to the occurrence of the critical rainfall which then results in flooding.

Such situations are expected to occur, and do occur, in cities of the Guinea coast in June and July. (For example, the storm of 15-16 July, 1970 in Lagos started at 20.00 hours on 15 July with low intensities until 22.00 when 63 mm (2.46") occurred within an hour. This was followed by a monsoonal type of rain of low intensity for the next 5 hours. Then as from 03.00 hours on 16 July moderate to very high intensities prevailed for 8 hours, flooding large areas of Lagos). It is the very frequent high-intensity short-duration thunder storms sometimes being superimposed on those of the monsoon type - causing extremely high runoff - that is the main source of the trouble. Another factor leading to flooding in these coastal plains is the low infiltration capacity of soils due to their colloidal nature or to the high ground water table which may almost reach ground surface.

Natural factors are of great significance in the generation of flood flows but man-made causes further aggravate the problem in towns and cities. In the development of urban areas man has not only concentrated many of his developments so that damages due to flooding of only a small area may be very expensive, but he has also disturbed the natural topography so that natural disposal of storm water is no longer possible. The unprecedented explosive physical growth of many African towns and cities during the last decade has replaced the former forests and farms with hectares of concrete, slate and bitumen and has created many complex problems for the present day planner and hydrologist as well as drainage engineer. These impervious surfaces drain rapidly into pipes and channels, concentrating rainfall,which results in flood flows the drainage channels can no longer contain. In urban areas with few or restricted natural or man-made channels sheet flows flooding roads and property may be a regular occurrence.

The flooding problem is further aggravated by poor and improperly graded streets and poorly maintained drainage channels. Often the kerbs and drainage ditches are blocked by garbage and other obstacles which impede flow and cause water to back up onto the streets and adjacent properties. On the whole, the existing drainage network is completely inadequate to cope with the amount of water released during flood flows because construction of necessary drainage works has lagged behind the rapid urban growth and the resulting runoff. The lack of proper planning and land zoning within many African towns and cities and their environs has encouraged the rapid, haphazard construction of shacks and slums directly in the way of water courses without alternative provision for drainage. Water backs up and collects in ponds which overflow and create local flooding. Finally the lack of central water agencies responsible for co-ordinating planning and implementing projects on environmental protection has also contributed to the problem by delaying or even preventing improvements to urban drainage conditions.

Water Pollution

Water carries wastes, diluting and stabilising them both chemically and biologically. Solar radiation and oxygen are major elements in this process. The availability of oxygen is influenced by temperature and other weather conditions (WMO, 1964). The first aspect of water pollution therefore concerns water quality as measured by indices such as PH value, biochemical oxygen demand (BOD), dissolved oxygen, temperature, turbidity, coliform counts (Wood, 1967) and radio activity. The other aspect is that of aesthetic nuisance caused by the like of floating cans, bottles, oil slicks and other floating garbage. Objectionable odour could also be classified as an aesthetic nuisance.

Water quality and pollution abatement are necessary not only for preserving fish and wildlife for food and recreational use, but also for other reasons. One of the main functions of water ways in a metropolitan area is that of drainage after a storm. Usually there is insufficient water flowing through these channels to make them useful for any other purpose. This inadequate flow (especially during long dry seasons)is a chronic problem that usually tends to be overlooked. Like the flooding problem, water pollution has been accentuated by the existence of drainage ditches constructed many years ago across the natural slope of the land. This topography has since been disturbed and stagnant muddy pools of water breeding mosquitos, intermingled with dense vegetation growth choking the ditches, are another direct result of low flow.

It is sometimes difficult to convince the public of the need to keep these drainage ways clear. In many cases, however, little or no effort has been made to improve the public perception of the hazard involved in plugging the few available water ways. Yet unless the authorities educate the people in this direction and, if necessary, compel them by legislation, the problem will persist for a much longer time.

There is no doubt therefore that water pollution is a multidimensional problem and, as such, must be attacked from many fronts. A realistic solution while attempting to reduce pollution in urban areas, will have to come to terms with the fact that concentration of population and industrial activity in our urban centres will inevitably result in some polluted water. This view is further buttressed by the fact that African urbanisation is frequently accompanied by a proliferation of septic systems and sewage-treatment plants. With the construction of this multitude of waste-disposal facilities - especially septic tanks - seepage causes nutrient enrichment (eutrophication) and bacterial pollution of both surface water and ground water. Unfortunately the servicing of many of these systems is often inadequate.

The problem of water pollution is further complicated by the very high rates of rural-urban migration already mentioned. It so happens that there is a great disparity between the living conditions of rural and urban areas so that when the inhabitants of the former come to live in urban areas, the tendency is for them to inadvertently engage in practices of hygiene which are inappropriate to and incompatible with the crowded conditions of metropolitan areas. These former rural dwellers thus need to be helped to adapt to urban conditions through education and legislation aimed at curbing environmental pollution and other related problems.

Water pollution problems whether in rural or urban areas must be attacked on a basin-wide level to co-ordinate the efforts of the communities concerned since the action of one community can affect others living downstream of them. The problem of water pollution is not, as many people believe, non-existent in rural areas. We have seen that in many African villages there is no dependable water supply so that for their drinking water villagers have to rely on natural sources, mainly rivers and streams, which are also used for other purposes such as bathing (and recreational swimming), washing of clothes and occasionally for transporting sewage. Rural communities down stream have to rely on natural purification of their water supply which has become progressively polluted as it passes through more and more villages.

It is therefore obvious that urban pressures combined with a lack of public perception and a lack of sound land use planning are the major factors in the problem of water pollution in Africa. Water and land use problems are interwoven and it must be apparent that a comprehensive programme would be needed for the proper management of both in a watershed. This inter-relationship is perhaps most clearly illustrated by the problems of water and land pollution caused by the rising water-table in Egypt. Waterlogging in very humid areas under certain topographical and soil conditions is perhaps a familiar difficulty, but a rising water-table in arid and semi-arid regions where intensive irrigation is practised is not such a commonly publicised problem as far as Africa is concerned. However, in the West Nubarya area of Egypt measurements taken over a period of three years (September 1969 to September 1972) revealed that in places the water-table rose by an average of 4 m per year, and is now found locally at depths of no more than 3 m below the ground surface (Schulze and de Ridder, 1974). The flowing ground water is saline, so its discharge into irrigation canals contaminates the fresh water in the canals. In addition, a major portion of the ground water flows towards the low-lying areas of the deltaic plain and causes, or worsens, their waterlogging and salinity problems. Artificial surface and sub-surface drainage is therefore required to reclaim both water and land thus polluted to prevent further deterioration of the environment.

Proposed Solutions

In attempting to solve the problem of water shortage in Africa, great stress has been placed on the provision of more water for the larger and more rapidly growing urban and rural communities. This is done by adopting an extensive approach which involves going further afield to obtain additional supplies, constructing larger dams, or drilling deeper wells. Obviously, it is important to provide water in areas where it is lacking, but at the same time, improvements in the efficiency of its use need to be publicised, especially in view of the increasing costs of constructing water supply systems as the scale of operation increases. This section therefore first discusses conservation measures as an alternative in water resource management. Here every member of the community, be he an official, or just a consumer, has a part to play. Conservation of water resources involves a whole host of different adjustments and may be conveniently discussed under four main headings: control of evaporation loss from reservoirs, improved efficiency of distribution systems, the elimination or reduction of wasteful use of water and the recycling of renovated waste water.

Evaporation losses under most African climatic conditions are quite high. The annual loss from a free water surface ranges from some 600mm in the equatorial belt of constantly high relative humidity to over 1,800mm in the Sahelian zone, not to mention evaporation under desert conditions. The problem of evapotranspiration (evaporation and transpiration) loss is of particular importance in sub-humid and arid parts of Africa where atmospheric humidity is generally much lower and in addition temperatures are very high. If water is to be stored for use during the dry season and held over for much of this period (up to 7 months in many areas), much of it will be lost to the atmosphere.

Many months' water supply may in effect be lost from a large water supply system of storage dams, or reservoirs and distribution channels. Such losses, for example, were observed by the present writer for Jos (Nigeria) water supply systems located within a relatively humid tropical environment with about 6-months dry season (cf. Oyebande, 1973).

It is for this reason that it is considered profitable to try to reduce such water losses by employing evaporation suppression controls. One such control involves the use of mononuclear layering on storage reservoirs, lakes, or other stretches of water to suppress vapour emitted from such surfaces. Evaporation control measures used to be considered more theoretical than practical, but research workers like Mansfield (1955a & b; 1964) and Vines(1960) among others have successfully demonstrated the practicability of the method. Some field trials show that the process can reduce evaporation by from 15% to as much as 70% and the cost of operation is sufficiently low to make it profitable. In fact, evaporation reduction is already an economic method of producing domestic water in Australia. Africa, with similar water problems, can benefit immensely by adopting this technique where environmental conditions make its application profitable.

Substantial water losses also occur through leakage from the distribution systems when pipes burst or joints are bad. In the city of Chicago, for instance, some 230 million gallons per day were lost through leakage in 1961. This amounts to nearly 20% of the total withdrawal (Maclver, 1970). In many towns and cities of Africa, where water authorities can ill-afford such losses, leakage is already a very serious problem. Through efficiency adjustments such as vigorous inspection of pipelines using electronic tracking of leaks, replacing of older pipes and improved methods of jointing, such leakages could be considerably reduced.

The general attitude to water in Africa and in many parts of the world puts it in a unique position from the point of view of resource economics. Water is often regarded as a free good, or a cheap one which should not be taxed. This kind of myth or misunderstanding surrounding water has often led to its wasteful use. There is an urgent need to prove to people that water is not a free but a valuable and expensive good which must be conserved with great care. But mere education is not enough. Some other concrete steps must be taken. In many African countries the present method of charging for water is ineffective in bringing about its conservation. A more effective method is to charge for water by measuring consumption, by metering all supplies as is done for electricity. This may appear to be an expensive exercise, but perhaps the end will justify the means. Care must be taken however to ensure that the consumption of water is not restricted below a desirable minimum compatible with the rising standards of living.

The recycling of renovated waste water from households and industry is becoming a practical and permanent source of 'new' water although this is generally regarded as an emergency measure. Here on our continent, Windhoek in Namibia (Johnson, 1971) began using renovated waste water for one-third of its requirements in 1969. Other arid and semi-arid and even sub-humid parts of the continent may start thinking of adopting this adjustment, to supplement other measures at least.

It would thus appear that water authorities in Africa must begin to emphasise a more intensive approach - a more efficient use of what is already available - in their efforts to obtain new or additional supplies. The problems of flooding and pollution have also been discussed especially as they affect African towns and cities. There is a special need for a pattern of adjustment that calls for more than just control of floods and pollution. A necessary, if not sufficient, condition for choice of adjustment is the motivation to reduce these hazards. Such motivation can only come from perception (Burton, Kates and Snead, 1968) of the danger and a knowledge of the available means for hazard reduction. The authorities charged with water resource and environmental protection must not only become aware of the danger themselves but must be prepared to educate and, if necessary, compel the general public to co-operate in reducing hazards.

There is a need for a new strategy altogether in water resource management in Africa. The strategy so far employed in most African countries has consisted of withdrawing water from surface or other sources for use, the water after use being discharged downstream into a network of natural and man-made channels which lead the waste and storm water into bigger rivers, lagoons or direct to the sea. This narrow straight-line strategy is becoming difficult to operate in many of our cities as a result of steadily expanding urban development in which hundreds of hectares of farmlands and swamps are converted to urban use every year. The 'downstream' (Rockwell, 1968) for discharge of both waste water and flood water is disappearing fast as urbanisation sprawls from central areas.

This accelerating rate of urbanisation and steady population growth coupled with lack of sound land use planning has further complicated the water problem. In view of the dynamic nature of the water resource management process, the present strategy needs to be reviewed and modified to include (as we observed for Namibia) recycling concepts based on reclamation and re-use of water. Under a new strategy of this sort the solution to drainage and pollution problems becomes something more than mere disposal of waste or flood water, and is linked to the water supply problem in such a way that 'excess' drainage may be directed to water supply systems and used to cater for domestic, recreational and other purposes. Not until this strategy is fully utilised can we overcome the ironical and problematic situation in which water shortage and flooding occur simultaneously in the same parts of the African environment!

Optimum Utilisation of Water Resources: the Need for Comprehensive Scientific Planning.

Attempts by a number of African governments to solve their water resource problems have been discussed. It does not require much effort to see that such attempts are proceeding in an unco-ordinated manner. This is a serious situation which has resulted in extravagant and uneconomic use of a vital resource and may in the future hamper and even jeopardise economic development. For efficient and economical use of water resources there is a need for integrated and co-ordinated scientific planning for conservation, utilisation and control. Such plans have to be comprehensive and basin-wide to ensure fair distribution of water for various requirements, including irrigation, rural, urban and industrial water supply, hydro-power, navigation, fisheries and others. To achieve this goal each country will have to undertake a proper survey and assessment or inventory of its water resources on a national scale. A master plan for their control and utilisation should then be prepared.

Basic hydrological data along with aerial surveys provide essential material for research, and insight into problems that require research. Lack of such basic data has retarded progress in hydrological investigation in Africa. The existing hydrological and hydrometeorological networks in most parts of the continent are deficient. There is a need to improve these networks for data collection, and to standardise the methods of instrumentation, observation and data recording to curb the present multiplicity of standards which renders comparison of data obtained in different regions difficult, if not invalid. In addition the data collected needs to be processed and published while concurrent research is undertaken for improvement. These tasks call for technical and professional staff of different categories who are very often in short supply. Facilities are now available, however, in some African countries for training such staff in Colleges of Technology, Departments of Water Resource Technology and University Departments of Engineering, Geography and other environmental sciences.

Again, before comprehensive water resource planning and development can be successfully carried out the awareness of the need to plan must find expression in national policies. We have much to learn from the more developed nations of the world. Britain and the USA are good illustrations of the increased government involvement in water resource management which results from increasing demand for water and severe droughts. In Britain the drought of 1959 coupled with demand reported to be growing at the rate of 2.4% per year from 1955-60 (UK Central Advisory Water Committee, 1962), and 4% per year by 1970 (Bugler, 1971), stimulated a concerted national water policy embodied in the Water Resources Act of 1963 which at the same time set up a central water authority responsible for co-ordinating water resource conservation and management. In the same way the growing demand for water and a prolonged severe drought, which began in the north-east of the USA in August 1961 and continued unabated for 46 months, provided the stimulus for establishing an overall national water policy by speeding up the approval and implementation of already planned new water legislation (Hogarthy, 1971).

The continent of Africa is faced with similar problems of fast growing demand for water for all purposes and the recent severe drought in the Sahelian zone, presenting African countries with an opportunity to develop their water resources by establishing national policies (Oyebande, 1974a; 1974b). Such policies can be crystallised in legislation (or decrees) to set up a central Water Resources Authority in any particular country as an independent and supreme body for directing the centralised functions of data collection, processing and dissemination. Other functions of such a body would include division of the country into hydrological regions each comprising a number of river basins within which, in addition to basic hydrological data collection, special hydrological, hydrometeorological, and hydrogeological studies should be conducted as and when necessary. These studies may comprise storm rainfall-duration-intensity, depth-area-duration analysis, evaporation and evapotranspiration mapping, drought intensity and frequency, flood frequencies, runoff mapping, etc., all of which are very useful for the planning and design of water projects.

Conclusion

The water resource problems of Africa are complex and multifarious. A combination of many approaches is thus required in any attempt to solve them. The expansion of industries, population growth and rising standards of living will continue to create increasing demand for water. Such demand is bound to create local shortages of water in as much as the vast volume of the river Niger, for instance, cannot be made available in the Sahara desert. These local shortages will in turn remain a limiting factor to development in most arid and semi-arid parts of Africa. At the same time water pollution constitutes a great problem as it diminishes both the quality and quantity of available water resources by necessitating expensive treatment before such water can be safely used for certain purposes. Application of modern technology to water resource management, however, has enabled man to harvest water which used to be unavailable. For instance, the improved efficiency of benefication of saline water and the development of deep well turbines for probing deeper aquifers have made a resource out of water formerly locked up in reservoirs and isolated from water supply systems. Africa has benefited from such technological applications and may be expected to gain more in the near future.

As the water need in Africa expands, even with the application of conservation measures, the scale of operations of water systems will expand both in terms of the cost of construction and maintenance, and of their effect upon the environment. Consequently, decision making with regard to such projects becomes increasingly a matter of great responsibility which requires training and experience. Unfortunately, the relevant personnel is still lacking in African countries, but this problem should prove tractable if the development of water resources is organised in the way outlined above.

Furthermore, instead of the present series of unco-ordinated water resource projects which characterise many African economies, water resource development must form part of regional or national development plans supported by a conceptual framework, embodying a set of multi-purpose objectives. This is the way African governments can play their proper role of shouldering increasing responsibility for maintaining the rate and direction of growth in expanding and dynamic societies. Regional economic cooperation has already brought a number of African nations together and should be extended to water resource management. Effective water planning and development in Africa in fact will necessitate such cooperation. For instance, Lake Chad has four countries (Cameroun, Niger, Nigeria and Chad) sharing its shores. The rather high population growth of this area gives an added urgency to the need to adopt a collaborative development plan for a region where the dry season lasts at least seven months in any year. Because it was recognised that progress in this part of the Sahelian zone depends on the full utilisation of its water resources, the Chad Basin Commission, comprising the four countries, was formed in May 1964 to study and develop the water resources of an area of some 40,000 km, in cooperation first with UNESCO (cf. UNESCO, 1970) and then FAO.

From the foregoing it is obvious that, in Africa, there is an accelerating recognition that water is often the key to industrial prosperity and agricultural productivity. The national policy on scientific water research aimed at solving the water problems which must arise in a dynamic society is therefore very important in the political climate of all African countries. For, in spite of the current mineral boom which some of their economies are experiencing, all African nations will for many years to come need increased agricultural production, not only for domestic food supplies but for earning foreign exchange. Optimum utilisation of water resources which, as we have seen, are slender over large areas is essential therefore and will continue to be a major focus for agricultural and hydrological research in the African environment.

REFERENCES

BUGLER, J. 1971 'Will Britain run dry?', *Observer*, 16 May. London.

BURTON, I.R., KATES, R.W. and SNEAD, R.E. 1968. *The Human Ecology of Coastal Flood Hazard in Megalopolis*. University of Chicago Department of Geography Research Paper no. 115 . 196p.

DE GREGORI, T.R. 1969 *Technology and the Economic Development of the Tropical African Frontier*. Cleveland and London.

ECA 1971 *Survey of Economic Conditions in Africa, 1971*. Part 1. Economic Commission for Africa, Addis Ababa and New York.

GHANA PLANNING COMMISSION 1964 *Seven Year Development Plan of Ghana 1964-70*. Accra.

HOGARTHY, R.A. 1971 'Metropolis runs dry', *Journal of the Irrigation and Drainage Division, American Society of Civil Engineers (ASCE)*, 97, IR4, 559-70.

ISAAC, C.G. 1965 'Water, waste and wealth'. Inaugural lecture, University of Newcastle-upon-Tyne.

JOHNSON, J.F. 1971 *Renovated Waste Water as an Alternative Source for Municipal Water Supply*. University of Chicago Department of Geography Research Paper no. 135. 115p.

KLASSE-BOS, A. 1969 'Rural Water Supply in Kenya: its role and development', in: D. Warner (ed.), *Rural Water Supply in East Africa*. Proceedings of workshop at University College, Dar es Salaam. (BRALUP Research Paper no.11).

MACIVER, I. 1970 *Urban Water Supply Alternatives*. University of Chicago Department of Geography Research Paper No. 126 .

MANSFIELD, W.W. 1955a *Summary of Field Trials of the Use of Cetyl Alcohol during the Season 1954-5*. Commonwealth Scientific and Industrial Organisation (CSIRO) Division of Agricultural Chemistry Report no. 174.

—— 1955b 'Influence of monolayers on the natural rate of evaporation of water', *Nature*, 175, 247-9.

—— 1964 'Control of evaporation', in: Australian Academy of Science, *Water Resources Use and Management*. Proceedings of a symposium held at Canberra 9-13 September 1963. Melbourne. 112-15.

NIGERIA FEDERAL MINISTRY OF INFORMATION 1970 *Second National Development Plan of Nigeria 1970-74*. Lagos.

OYEBANDE, L. 1973 *Some Aspects of the Hydrology of Urban Water Supply - a Case Study of Jos.* Paper presented at the Nigerian Geographical Association Conference, December 1973. (Proceedings in press).

—— 1974a 'Water conservation in economic planning', *Daily Times of Nigeria,* 24 May.

—— 1974b ***Water*** *Resources Development and National Welfare.* Lecture given to the Nigerian Field Society, University of Lagos, College of Education.

—— 1974c 'Water resource planning management', *New Nigerian* 14 June.

—— 1974d 'We must conserve our water resources', *Daily Times of Nigeria* 11 April.

—— 1974e *Storm Hydrology and Problems of Flood Plain Development in Greater Lagos.* Paper presented at the Nigerian Geographical Association Conference, University of Nigeria, Nsukka.

ROCKWELL, M.L. 1968 'Water resources as an element of urban planning', *Journal of the Urban Planning and Development Division (ASCE),* 94,NP1,1-10.

SCHULZE, F.E. and DE RIDDER, N.A. 1974 'The rising water table in the West Nubarya of Egypt', *Nature and Resources (UNESCO),* IX, 12-18.

SZENTES, Thomas. 1968. *Introduction to the Economy of Tropical Africa.* Report of the Centre for Afro-Asian Research, Hungarian Academy of Sciences, Budapest.

UK CENTRAL ADVISORY WATER COMMITTEE. 1962 *Subcommittee on the Growing Demand for Water. Final Report.* HMSO, London.

UNESCO. 1970 'Study of water resources in the Chad basin', *Bulletin of the International Hydrological Decade,* VI, 6-9.

VINES, R.G. 1960 'Evaporation control', *Australian Journal of Applied Science* 11, 200.

WARNER, D. 1969 *Rural Water Supply and Development: a Comparison of Nine Villages in Tanzania.* Dar es Salaam University College, Economic Research Bureau Paper no. 69. 17.

—— 1970 *The Economics of Rural Water Supply in Tanzania.* Dar es Salaam University College, Economic Research Paper no. 70.19.

WOLF, P.O. 1966 'Comparison of methods of flood estimation', in: Institution of Civil Engineers, *Proceedings of a Symposium on River Flood Hydrology.* A1. 1-23.

WOOD, D.F. 1967 'Water front renewal in metropolitan areas', *Journal of the Urban Planning and Development Division (ASCE),* 93, UP4, 141-150.

WMO. 1964 *Weather and Man.* World Meteorological Organisation no.143 TP 67. Geneva.

WATER AND ENVIRONMENT IN TANZANIA*

LENNART ANNERSTEN
Uppsala University

The question of water supply is probably one of the main issues in the struggle to achieve better living standards and economic development in East Africa. Water supply has certainly been a main concern in these countries and its social and economic importance is well established. By East African standards large investments in water works have also been made to cope with a growing demand for sufficient clean water. In 1973, Kenya for example earmarked approximately £4 million for rural and urban water supply, emphasising the rural areas. In Tanzania still larger efforts have been made and in its current 5-year plan approximately £28 million will be spent on water supply.

Marginal Efforts

The financing of these water supply schemes is met to a large extent by external aid. For example water supply projects are amongst the largest within the Swedish aid programme, representing roughly 7-8% of Sweden's total aid.

On the other hand it must be remembered that by West European standards this contribution is rather modest. Investment costs in water supply and sewage facilities in Sweden itself amount to approximately £150 million per year. Even if a direct comparison between Swedish and East African conditions in some aspects is inappropriate, it is quite obvious that aid spent on water development, from the industrialised countries' point of view, constitutes hardly more than a marginal effort.

There is no doubt, however, that if these aid programmes are fulfilled and comply with targets set forth in the development plans, they will bring great social change, especially for the rural population. By means of these changes it is hoped to facilitate economic development, introduce better health conditions, give more time for productive work and implement better conditions for agriculture and ranching.

It must be stressed, however, that these water supply schemes do not aim to provide the East African citizen with vastly increased quantities of water. The capacity of the schemes is based on an estimated consumption of 10 gallons (45 l.) per person per day. By comparison a West European city consumes approximately 100 gallons (450 l.) per person per day. It has been a major political concern to supply sufficient clean water to most areas in East African countries. There has also been some attempt to evaluate the effect of such political decisions by measuring the rate of change of economic, social and health levels. The importance of such evaluation is also quite obvious.

What is lacking in the general evaluation is the hydrological foundation for carrying out the programmes and an assessment of the hydrological resources available for an expanded programme. A political decision does not automatically mean that the necessary water resources are available.

Unknown Water Resources

What are the hydrological conditions for an extended water supply scheme in Tanzania? With the knowledge we have today it is impossible even to estimate the potential water resources which could be developed. Reliable information about available water resources only exists for certain very limited areas, particularly larger rivers such as the Pangani, Wami, Rufiji and some smaller drainage basins. These areas, however, represent a very small part of Tanzania's total land area. As far as we know today, in most areas, Tanzania must rely on ground water supplies, not surface water.

The Water Balance - A Critical Factor

The water balance is the difference between water income and water loss. The water income is mainly precipitation, together with possible drainage into an area. The water losses are evaporation, water consumption by plants etc., and discharge from rivers.

As a rule, northern Europe has a positive water balance, mainly due to low evaporation. In Sweden, for example, an average annual precipitation is about 700 mm, average evaporation about 350 mm. There is a surplus of water which results in a constant and reliable discharge in rivers and a natural refilling of ground water reservoirs. As long as this natural water balance is maintained the storage of water in the natural reservoirs will be constant. If artificial withdrawals of water from these reservoirs are too large the system will be affected in such a way that discharge from the rivers is reduced and the ground water table sinks. This has already happened in highly industrialised areas in northern Europe and North America and some arid areas of the USA.

Tanzania, on the other hand, has a negative water balance. Typical precipitation figures for the central parts of Tanzania are 500 - 750 mm/year, while for eastern and western regions they are slightly higher, at 750 -1250 mm/year. Only 4% of Tanzania's land area can expect with any degree of confidence a precipitation of more than 1250 mm/year. For comparison the potential evaporation in Tanzania is about 2000 - 2200 mm/year.

Potential evaporation - or evaporative demand - is the evaporation from an extended open water surface, e.g. a dam. Available data indicate that the water deficit for the main part of Tanzania amounts to approximately 1000 mm/year or about 1 m^3 per m^2 and year. Even if that figure is somewhat hypothetical, it clearly indicates that a water surplus is hardly likely to develop and that shortage of water and water stress must be the normal condition for whatever life there is.

It may be suggested that areas with rainfull of less than 750 mm/ year are marginal from the agricultural point of view, bearing in mind that seasonal distribution is important and makes it impossible to define clearly a lower value. The table indicates, however, that only half of Tanzania fulfils this requirement even provided that one is ready to accept a drought every fifth year! Only 1/5 of Tanzania has sufficient rain to meet the standard of a drought one year in ten.

These figures give clear indication that, from the hydrological point of view, there are very small margins for agriculture. The question of famine or plenty in Tanzania is in this respect a matter of millimetres.

Unknown Ground Water Reserves

The strong negative water balance in Tanzania will, in addition to creating a general shortage of surface water, also affect the accumulation of ground water. Ground water is formed in two ways - by infiltration of rainwater from the surface, or leakage from rivers and lakes. Our knowledge about the ground water reserves in Tanzania is indeed very scanty. Reliable estimates only exist for very limited areas and even for those the recharge is practically unknown. What is said below must thus mainly be guesswork based on general hydrological conditions, rather than knowledge based on facts.

The occurrence of larger ground water reservoirs is favoured by special geological structures. In arid areas, these are generally associated with sedimentary rocks. In Tanzania, there are very few areas of sedimentary rocks - probably less than 10% of the basement in Tanzania is overlain by them.

Water bearing structures and aquifers can also develop in volcanic areas. Such areas do exist around Mount Kilimanjaro, for example, but the water is very likely to be alkaline and of limited value. This is often the case in the Rift Valley, which in itself has a favourable structure for storing large ground water reserves.

Crystalline rocks, mainly granites and gneisses, are the dominant rocks in Tanzania. Locally, ground water reserves that occur in these rocks could be of importance and are in such cases associated with cracks and deep weathering. In practice, it is very difficult to find these reservoirs. Exploration of these ground water reserves is therefore both expensive and unreliable.

The existence of large ground water reservoirs, comparable with those found in parts of the Sahara and the Indus valley, are practically out of the question in Tanzania. Only the coastal region, Rufiji-Luwego-river valley, and a narrow zone in western Tanzania have, from the hydrological point of view, favourable conditions for large continuous ground water reserves. There have been no investigations, however, giving the necessary data for a more accurate calculation of the potential water resources of these areas.

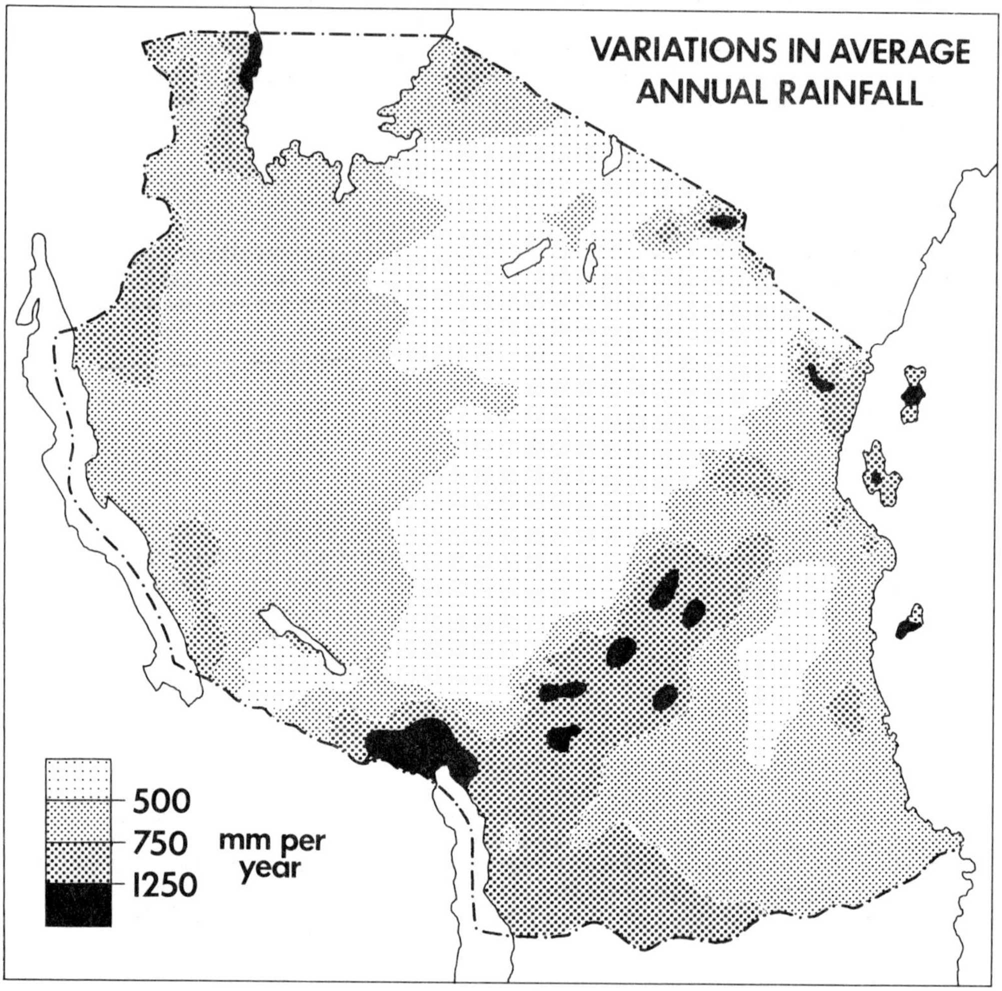

Fig. 1 Rainfall in Tanzania

Unfavourable Distribution of Rainfall

The reason why crops can be grown in spite of these conditions is, of course, the seasonal distribution of rain even if the rainy season is, as a rule, shorter than the dry season.

Large areas of Tanzania have only four months with rainfall in excess of 100 mm and none above 150 mm, succeeded by a seven months dry period in some areas. The western and eastern regions have a rainy season of about six months and a marked dry period of about four months. The characteristic seasonal oscillation between a dry and wet period develops woodland and savanna as climax vegetation, which in Tanzania has probably reached its most extreme form.

In northern Europe the precipitation pattern is determined by the frequency of cyclones passing over the continent. On the average, these amount to about seventy per year. Rainfall variability is in fact rather small, even if drought periods do occur locally. Statistically, only one year in a hundred will receive a rainfall $\frac{2}{3}$ of the normal.

In East Africa deviations of this magnitude are much more frequent. Precipitation $\frac{2}{3}$ of normal occurs on average one year in five, or even more frequently. On the other hand, excessive rainfalls can occur with about the same frequency. The surplus of water from excessive rain will mainly result in a larger discharge from rivers and cannot be utilised. In practice, it will be the periods of extreme drought which set the limit for agriculture. Available water resources are in effect determined by the amount of water that can be mobilised during these frequently recurring dry years.

It might be of interest to see how the 'normal rainfall' is distributed over Tanzania. The term 'normal rainfall' can be defined in ways, but from the agricultural point of view it is desirable that at least four years out of five (80% probability) or, better still, nine years out of ten (90% probability) have enough rainfall to secure a proper harvest. The table below gives the figures for these conditions.

Precipitation in mm/year	Percentage of land area with 80% probability of receiving a given amount of rainfall	Percentage of land area with 90% probability of receiving a given amount of rainfall
300 mm	16%	32%
500- 750 mm	33%	47%
750-1250 mm	47%	18%
over 1250 mm	4%	3%
	100%	100%

Ground Water Refill- A Question-mark

In areas where more extensive ground water reserves are lacking this could, in principle, be compensated for by the combined effects of rapid recharge and discharge in the aquifer, and in increased turnover in the ground water reservoir.

Very little is known about ground water formation in Tanzania, either as a process or quantitatively. There are some indications, however, that inselberge and isolated mountains on the savanna plains are important infiltration points.

These isolated mountains usually have significantly higher precipitation than the surrounding areas, and the discharge from them very often results in ground water flows. This is probably the case in Mzima Springs in Kenya, for which the nearby Chulu Range is the most likely water source.

The very limited number of investigations that have been made, however, indicate that ground water formation is a very slow process and that a very small part of the rainfall actually infiltrates to the ground water table.

At Muguga, about 30 km north of Nairobi, relevant measurements were made during a 16-year period. Out of a total of 16,600 mm of rain only 375 mm reached the ground water table, i.e. about 2-3% of the actual precipitation. In fact recharge only occurred during two exceptionally rainy periods, while periods with normal rainfall added nothing at all.

Keeping in mind the negative water balance in East Africa it is not at all surprising that ground water formation should be such a slow process. The very few natural wells are certainly an indication that ground water reserves are limited and that the ground water table is situated fairly deep.

As long as there are no datings of the ground water it is impossible to estimate the rate of turnover in ground water reservoirs. It is not unlikely that much ground water is fossil, being formed during a period when other climatic conditions were prevailing. This is yet another of the large question-marks concerning Tanzania's hydrology.

Soil Erosion and Sedimentation

The climatic pattern with its alternating rainy and dry periods affects the vegetation. By the end of the dry period there is no continuous cover of the ground and when the rains start the impact of raindrops on the bare ground will cause soil erosion. Erosion problems are greatest in arid areas where occasional heavy rains probably cause most damage. Extreme rainfall in mountainous areas may also result in severe earthslides.

There is no single measure which would stop soil erosion. One prerequisite is that agricultural methods and activities are adapted to the natural conditions. A new social pattern might be necessary in certain areas; in other areas it could be equally important to preserve a social and cultural pattern which has developed over generations and is in harmony with the natural conditions. Any activities which are planned and implemented must, however, be founded on a quantitative knowledge of the hydrological and geological processes occurring in the area.

Soil erosion also means increased sediment transport in streams and rivers. There are very few investigations of sediment transport in East African rivers. In peak floods, however, sediment loads several hundred times larger than normal maximum values for Swedish rivers have been recorded. Concentrations of suspended load of the order of 10 kg/m^3 have been recorded.

Dam sites and water reservoirs act as traps for sediment and there are examples of larger dams which have been completely filled with sediment within 15 years. Such a short lifetime for a water reservoir hardly justifies costly investment in dams and irrigation work, even if there are pronounced social reasons for building them.

Hydrological Research Necessary

Water is not a resource that is used once and for all. Our 'consumption' of water is just one part of the hydrological cycle and under certain conditions water can be used several times before it, in one form or another, disappears from an area.

In order to develop water resources in East Africa we must have a better knowledge of the hydrological conditions. With few exceptions this knowledge is still too limited to allow any sort of long-term planning. Basically, the problem is that the tropical savanna represents a special environment with unique problems so that technical solutions from other areas cannot be applied automatically.

The solution to these problems depends on a fundamental re-orientation of viewpoint, towards the view that water is not a commodity in excess but rather a very scarce resource.

In northern Europe, the problem of getting water is more or less confined to the technical exercise of collecting it. In East Africa, it is a matter of the very existence of water and, in the event of there being water, how long supplies will last. Technical solutions must be founded on a comprehensive knowledge of hydrological conditions.

Hydrological research and development is a long term activity, however, simply because long records covering several decades must be available in order to permit the right conclusions to be drawn. It is nevertheless a prerequisite for solving social and economic problems in these areas. Without that knowledge there is a risk that economic development may stagnate and that efforts being expended today will simply create a small breathing-space and will soon be overtaken by a rapidly growing population.

Fig. 1 EQUILIBRIUM IN THE PHYSICAL ENVIRONMENT

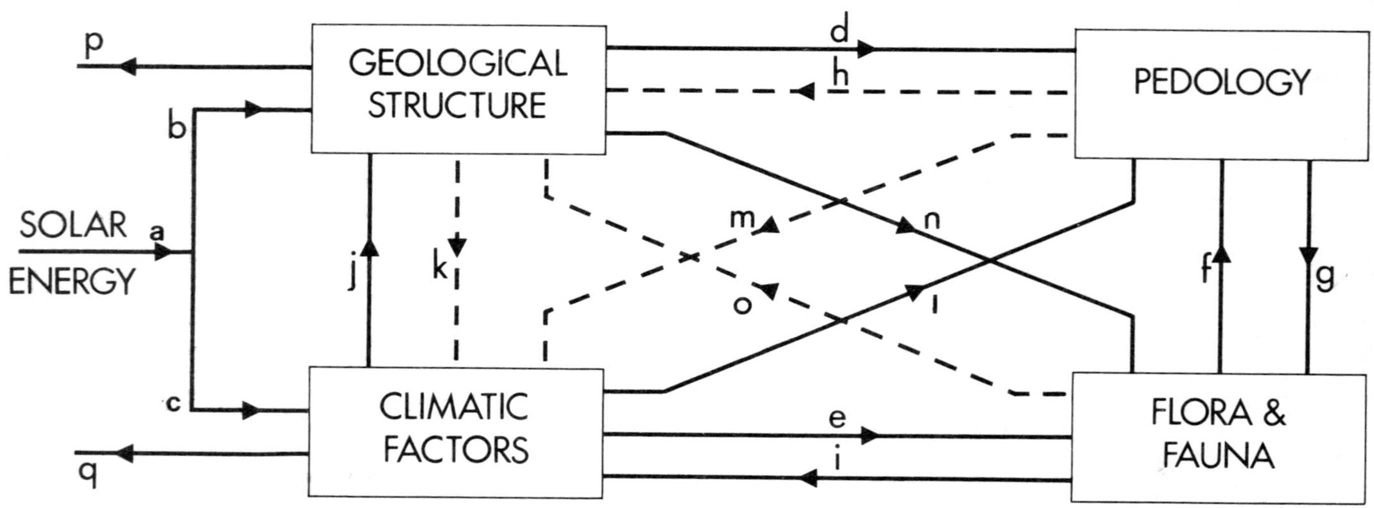

THE PROCESS OF SOIL EROSION*

PHIL O'KEEFE
Bradford University

The path of human development is best defined as an ecological process in which society has increased its capacity for controlling and utilising environmental resources. This capacity for controlling and utilising resources is dependent on the manner in which society is organised (the socio-economic and political considerations), upon the extent to which people understand the laws of nature (science) and upon the extent to which they put that understanding into practice (technology). If one is looking at agriculture in terms of such a definition then it is also necessary to consider the wider context of ecological equilibrium within the physical environment. This equilibrium is described in Figure 1.

The initial input is solar energy (a) which directly influences the geological and geomorphic structures and the climatic factors, (b) and (c) respectively. Geological and geomorphic structures directly influence the pedology of an area (d), in that the soil is weathered rock and its deposition depends on the varying geomorphological conditions. In a similar way, the climatic regime is linked to the flora and fauna of an area (e), usually providing a suitable environment for growth. Linkage between the soil and plant-animal life is a direct process in which certain types of organic life will prosper in certain soils and in which the life cycles will allow for the recycling and rejuvenation of that soil, (f) and (g). There is a weak link between the pedology and the geomorphology (h), in that soil structure could influence the geomorphic structures. A stronger linkage exists between vegetation and climatic structures particularly in terms of transpiration. Both energy flows (h) and (i), the latter delimiting the linkage just described between the vegetation and the climate, are particularly observable on the microscale. There is a strong relationship between the climate and geomorphological structures (j), particularly if one interprets the model in terms of geological time; a weak relationship exists in reverse (k), which again is observable on the microscale, e.g. orographic rainfall. Climatic factors have a strong influence on the soil, for it is the chemical processes in soil, consequent upon the climate, that largely weather the rock (1). There is a weak relationship between the pedology and the climate, especially in terms of evaporation (m). The influence of the geological and geomorphic structures on vegetation is largely indirect although there are distinct relationships, such as aspect, that play an important role in plant growth (n). The reverse relationship, the influence of vegetation upon the geomorphic structure, is weak except where removal of vegetation allows the development of soil erosion (o). Flows (p) and (q) complete the model and indicate that the system is in equilibrium.

*English original.

Man's ability to control and utilise this environmental potential rests almost entirely on his understanding of the vegetational complex. Since the neolithic revolution which brought the domestication of plants, it is utilisation of the vegetational complex that has enabled man to begin to control the environment. The process, however, is not simple because man's influence on the vegetational complex has repercussions throughout the total ecological system. First, the artificial establishment of a different range of vegetation will create a different demand for soil nutrients but second, and most important, the removal of vegetation facilitates soil erosion because of the absence of root material to hold the soil. Such erosion will ultimately affect the geomorphology of an area and subsequently the microclimate. In fact, in areas of severe soil erosion, man can be viewed as the major geomorphic agency. It is evident, therefore, that soil erosion is not an isolated physical phenomenon but is tied into the physical complex. The causes of soil erosion, however, are seldom physical but are largely social in origin, a consideration of paramount importance in land use planning.

The African environment suffers from soil erosion in both semi-arid and humid areas. Semi-arid areas receive most attention because in these areas productivity is falling, the land resource base is being eroded, and the population is living an increasingly precarious existence; discussions of desertification, however, tend to mystify rather than clarify the nature of soil erosion. Although the total human population of the semi-arid areas is not great, the population number vis-à-vis available resources is great. Pastoralism, with associated free grazing rights, is essentially an adaptive but invasive activity for, without control, it allows freedom of movement and freedom of herd size. Herd size is often increasing, creating a stable minimum state of vegetation; any further decrease in the vegetation mass or any increase in herd numbers encourages malthusian checks on the population. The plants, usually hardy annuals, frequently do not mature but are grazed as young shoots so that vegetation cover is rapidly removed. If, as in recent years, drought is prevalent, the vegetation is further removed, encouraging the removal of the soil itself by wind erosion.

It is obvious that major climatic changes and even short-term fluctuations influence soil erosion in semi-arid areas, but these changes only have such an effect because the population of the area cannot change its strategy of resource utilisation[1]. Before the colonial period, many pastoralists would have been able to pursue new strategies when the supply of pasture or the size of the herds changed but now fewer alternatives exist. A larger population is concentrated on the available resources, encouraging erosion, particularly around dry season wells and on the trails to urban markets.

To check such soil erosion, an integrated planning approach must be introduced incorporating measures that will check both wind and possibly water erosion and also conserve moisture; single factor development programmes cannot provide this integrated approach, consequently few enjoy success and some court disaster. An integrated programme would necessarily involve government legislation introducing statutory land legislation; any change in land tenure practice from customary to statutory status would have far-reaching ramifications in terms of social and political organisation, but it is only with such legislation that a balanced ranching system can be introduced wherein the ecological potential of the semi-arid areas is developed in such a way that it harmonises with the needs and value systems of the users.

[1] See P. O'Keefe and B. Wisner, in this volume p. 31

Soil erosion in the humid areas can be explained similarly in terms of human intervention. The removal of vegetation encourages sheet erosion on fine soils because the rain energy is not dissipated. Sheet erosion allows for the development of rills and gullies and eventually leads to the creation of 'Bad Lands'. The rate of erosion depends on the permeability of the soil and the infiltration rate of water but prevention can be attempted with terracing or bunding, especially on slopes, which reduces run-off velocity. The prevention of soil erosion is perhaps best seen as a process that attempts to preserve the crumb structure by maintaining humus content and by avoiding excessive cultivation which may, depending on climatic conditions, powder or puddle the soil. Bunding, a technique practised in Africa possibly since the neolithic revolution, not only prevents erosion but preserves soil moisture content and seedlings. It does, however, require a large labour input at a time when the decreasing importance of specialised roles in subsistence agriculture and the drift to urban labour have removed many males from the manpower pool. Lack of economic activity in the African urban industrial complex means greater pressure on indigenous agricultural resources. The growing population freed from mortality control but with no concommitant birth control removes residue vegetation as land hunger grows. Little rotation is practised and fragmentation hinders responsible collective action. Urban drift continues and the land quality deteriorates.

Although soil erosion is a catholic phenomenon, there is no preventative panacea. Restoration of land quality is a long term project because the kind of poor husbandry that destroys the catena destroys centuries of biological activity. It is not usually a lack of knowledge among peasant cultivators that encourages soil erosion,for their ethnoscience is rich, but rather the manner in which they are constrained to follow a single pattern of subsistence agriculture. A policy of radical conservatism is required which seeks to isolate the sociological causes of soil erosion. If this course is not pursued, man's relationship with the environment will no longer be commensal but predatory and much of Africa could become *resourceless* in terms of agriculture.

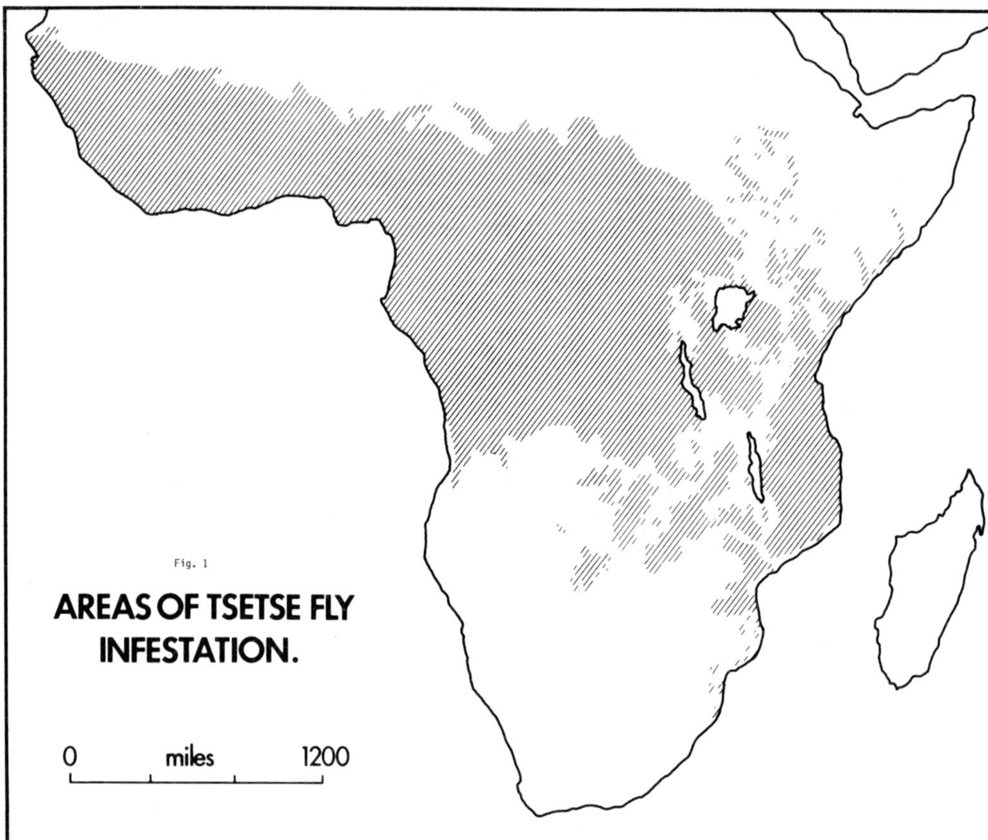

Fig. 1

AREAS OF TSETSE FLY INFESTATION.

AFRICAN TRYPANOSOMIASIS: AN ASSESSMENT OF THE TSETSE-FLY PROBLEM TODAY*

JOHN FORD

Sleeping sickness was once the most dreaded of endemic African diseases. From about 1890 to about 1935 a series of epidemics caused the deaths of millions of Africans. The closely related infections of domestic livestock were held, by some observers, to have had an even more disastrous effect, over the centuries, on the development of the continent, for they prevented the agricultural exploitation of vast areas south of the Sahara. This review attempts to indicate some aspects of these diseases in Africa today. It does not describe any of the extensive research programmes now being carried out in Europe and America.

In the decade before the attainment of political independence colonial officials, reviewing their own research and over half a century of practical experience, tended to say that the problems presented by the presence of tsetse-flies in Africa were virtually solved and that only time and money were required to deal with them. They were wrong.

The trypanosomiases are diseases of people and of domestic livestock caused by protozoan blood parasites called trypanosomes which are transmitted from one vertebrate host to another by the bites of blood-sucking insect vectors. In tropical America the trypanosomes of the widespread and dangerous Chagas' disease are transmitted by reduvid bugs. In tropical Africa tsetse-flies (*Glossina*) are the vectors of several species of pathogenic trypanosomes living in the blood of the larger wild animals, especially the antelopes and pigs, in which they are harmless. (Lion cubs and perhaps the young of some antelopes may become ill when first infected and even die before they have built up a natural immunity.) The commoner tsetse-borne trypanosomes are classified into three main species. *Trypanosoma vivax*, *T. congolense* and *T. brucei*. Domestic animals become infected by them when bitten by infected tsetse-flies. The first two named cause severe illness in cattle, usually ending, if untreated, in death. Sheep, goats and pigs may be similarly affected. *Trypanosoma brucei* is generally harmless to cattle, but kills dogs and horses. It has also given rise to two sub-species which infect man, causing diseases which are nearly always fatal unless treated. *T. vivax* and *T. congolense* do not infect man, nor do other strains of *T. brucei*.

Tsetse-flies are found throughout most of tropical Africa. They feed only upon the blood of larger vertebrates and do not occur where there is little tree growth or where the population of potential host animals is too sparse. On the equator they are not found, save accidentally, over 2000 m above sea level. The altitude limit decreases

*English original.

as distance from the equator increases, but as the tropics are approached zones with prolonged high or low mean temperatures provide environments in which these insects cannot live. It is thought that about 10,000,000 km^2 (4,000,000 square miles) of tropical Africa between the northern and southern deserts are infested with tsetse-flies. There are 30 species and sub-species of these insects and they fall into three groups. The *fusca* group is associated, for the most part, with the humid forests of the Zaire basin and of West Africa; the *morsitans* group includes most of the species found in the drier tree savanna and deciduous woodlands of east and south-central Africa, as well as the Sudan and Guinea zones of West Africa. Lastly the *palpalis* group is almost entirely confined to the wooded borders of lakes and rivers draining into the Mediterranean and the Atlantic Ocean and some internal basins such as Chad and Rudolf, but never those entering the Indian Ocean. This geographical fact is largely responsible for the marked differences in the epidemiology of the human trypanosomiases in West and East Africa.

Control over the human trypanosomiases had been achieved between 1930 and 1950 by the use of curative and prophylactic drugs supplemented in some areas by resettlement of populations at risk as well as by reduction or elimination of local tsetse infestations, usually by bush clearing. Until 1950 the only practical method of control of cattle trypanosomiasis was by control of tsetse-flies, either by bush clearing or by large-scale slaughter of their natural hosts, the wild animals. Towards the end of the 1940s, however, easily administered drugs for the cure of infected cattle began to appear, some of them showing marked prophylactic properties. At about the same time first DDT and then other insecticides like dieldrin began to be used in various ways to control *Glossina*.

In the generally optimistic picture seen by colonial scientists at the time of attainment of political independence only two technical flaws were apparent. The first was the possibility that improper or careless use of drugs to treat cattle might lead to the formation of strains of drug-resistant trypanosomes. The second was the danger of insecticide toxins building up in people, whether those engaged in control operations or merely living in affected areas, and in wild life. Such considerations led to a number of applied research projects which produced more or less satisfactory but not entirely complete answers to the problem.

Drug regimes were devised in which one drug was alternated, every few months, with another to discourage build-up of trypanosome strains resistant to either, while a third drug was held in reserve which could kill any drug-fast strains which might escape the normal routines. (Pharmaceutical research also led to the development of drugs which could cure human sufferers in the late stages of sleeping sickness which had hitherto resisted treatment). Dangers in the use of organic pesticides were diminished as a result of studies on tsetse-fly behaviour which showed that, in some bush types, spraying could be confined to limited portions of certain trees growing in plant communities which only formed a small part of the whole infested area. Where such discrimination was not possible, as in aerial spraying, new formulations, not only of insecticides themselves, but also of emulsifying agents, and new methods of spraying and fogging resulted in considerable improvements in efficiency. Thus in 1970 Rhodesia reported the successful selective use of 166½ tons of 75% DDT at an average rate of 1.8 oz. per acre, a rate which contrasted very favourably with insecticide applications to crop foliage of 1 lb to $1\frac{1}{2}$ lb per acre several times a season. At such low dosages the toxic effects on wild life are trivial, a conclusion also reached in a WHO study of deposits on vegetation and in tissues of wild animals in Kenya.

Stated thus, the achievements of applied research seem to have justified the optimistic assessments sometimes made at the end of the colonial era. It was a view readily accepted by politicians after independence who, anxious to promote investment, especially in tourism, tended to proclaim that the tsetse-fly problem had been solved. Some workers, especially those in human medicine, qualified their optimism by stressing the need for continuing maintenance, by the new regimes, of efficient control services. Thus a review of the impressive results obtained in the then Belgian Congo showed that incidence of sleeping sickness had been reduced to the low level, in 1958, of 0.019% of new cases among $6\frac{1}{4}$ million people at risk and under surveillance out of a national total of 13 million. The review concluded that the problem had become one of execution rather than research. What was required was an efficient and disciplined staff and the voluntary acceptance by the vast number of those at risk of periodical examination and treatment with curative and prophylactic injections, routines which had up to then been compulsory.

The review was published in 1963. By then the collapse of the control service was apparent. In 1959, just before independence, about 20% (i.e., some $1\frac{1}{4}$ million people) refused to submit to examination and treatment. In 1960 surveillance virtually ceased and only about 1% of the population in the endemic areas could be examined. Only a skeleton service was possible around the capital, Kinshasa. Then, suddenly, in 1964, it became apparent that the incidence of infection had risen to 6-10% with local foci showing 14-18% of the population infected. Thus there had been a 400-fold increase in the incidence of infection since the last accurate measurement in 1958.

It cannot be claimed that the epidemic is under control nor can there be, in the absence of adequate investigation, any agreement on its cause. One focus of severe infection studied by a WHO team lay about 100 km south of Kinshasa on the Angola border. The first major bid for Angolan independence, suppressed with great brutality in 1961, resulted in deaths estimated at between 10,000 and 30,000, while over 150,000 people fled across the frontier into Zaire. Some of these people were among the worst sufferers in the epidemic which broke out three years later. But the refugees could not be blamed for having imported the disease. They had come from parts of Angola where Portuguese services had operated a surveillance and control scheme very like that of the Belgians and using the same drugs. Moreover the infection was just as virulent in other areas where refugees were few.

A little later it was reported that refugees coming south into Uganda to escape from the violent disorders in the southern provinces of the Sudan Republic included some sufferers from sleeping sickness. It is now known that epidemics of great severity are afflicting the inhabitants of the region. In a number of features this outbreak resembles that on the Angolan border. Both were preceded by prolonged violence and in both there has been a high proportion of cases which do not respond to the drugs in common use.

The outbreaks in the Zaire basin are caused by *Trypanosoma gambiense*. In the Sudan both human trypanosomes are present. *T. gambiense* predominates west of the Nile, while *T. rhodesiense* infects the people of the drier eastern part. It has also been found in the valleys of western Ethiopia, a country in which human trypanosomiasis was unknown prior to 1967. The countries of eastern Africa, from Ethiopia southward to the Zambezi do not, expecially south of the equator, lie within the zone infested with *palpalis* group tsetses. To a large extent their infections are transmitted by flies of the *morsitans* group. There have been no violently disruptive epidemics but information available up to 1972 indicates a fairly steady increase in the number of reported cases of Rhodesian sleeping sickness. They are to be measured in hundreds rather than in thousands, but the general trend is disturbing.

In Africa west of and including Nigeria there does not seem to have been any deterioration in recent years. In 1967, however, a Nigerian medical survey team visited the population of an area in which there had been an epidemic of unknown origin, perhaps not of trypanosomiasis, in which all the sufferers had died.

The trypanosomiases, like many other diseases, are best thought of as manifestations of maladjustment in ecological systems. The details are complex and often obscure. One may consider the African trypanosomiases as involving five groups of organisms: the trypanosomes, the wild animals in which they are parasitic, the tsetse-flies which transmit them from one vertebrate to another and of which the wild animals are the natural hosts, and two groups of adventitious hosts, people and domestic livestock. By 1950 much was known about tsetse-flies and trypanosomes, but little, save in some aspects of pathology and therapy, about man and livestock. Of the interactions of the wild life in this system virtually nothing was known.

The control over these infections achieved by colonial scientists owed as much to control over human behaviour as to control over the tsetses and the trypanosomes. The latter could not be achieved without the former, but this was ignored in thought about the biological theory of these diseases. Only in settlement schemes designed to keep men and animals out of contact with tsetses did the interactions of vectors and adventitious hosts play any part. Compulsory submission to medical examination and treatment, confinement of domicile to circumscribed areas, limitation of movement and the more or less compulsory turn-outs of labour were regarded not as biological events but as conveniences to facilitate the work of the government technical services.

In the realm of cattle trypanosomiasis our predictions before 1960 also seem to have been inadequate. The fear that misuse of trypanocidal drugs would lead, through creation of drug-resistant strains of trypanosomes, to large scale mortality has not yet been realised. Drug-fastness has, indeed, been observed throughout the continent, but there have been no unusual large scale epizootics among the herds owned by nomad pastoralists or by peasant farmers. Among these people the standards of success, like the methods of husbandry, differ markedly from those characterising projects founded upon economic principles derived from Europe. The African cattle owner, perceiving failures among his animals, moves them to other pastures in the expectation, often realised, that they will escape the ills that beset them. Such remedies are not available to the manager of a ranch designed, as many externally financed schemes are, to occupy a circumscribed area.

The doctrine that the tsetse-borne diseases serve a valuable purpose in preserving the land from the effects of overstocking, was developed long ago. If this is a genuine problem, no answer to it has yet been found. A World Bank report on Nigerian development, published recently. notes that 'rangeland is insufficient and is deteriorating, due in large part to uncontrolled grazing by nomadic herds and encroachment of cultivation'(World Bank, 1974). The same report notes that it is unlikely that more than 40,000 square miles will be cleared of tsetse in Nigeria by 1980 and that this represents less than 60% of the estimated need for new grazing land which will be required to satisfy the projected needs of the local meat market. The Nigerian achievement in tsetse elimination far exceeds that of any other country, the runners up being Rhodesia and Uganda. One has to note, however, that in Nigeria eradication has, in some degree, been paralleled by the extension of fly-belts in other directions. And it is difficult to suppose that, in the other countries mentioned, recent events have not affected the level of efficiency required to prevent reinfestation of cleared land. In Africa as a whole one estimates that, since large scale attempts to eliminate tsetse-fly were begun in about 1925, total success may have reached about 1.5% of the infested area, but that almost certainly this has been more than balanced by reinfestation of formerly fly-free land.

In the light of the recent catastrophic Sahel famine the need to understand the human ecology of the African tropics has become very urgent. It is difficult to escape the conclusion that a major contributory factor in this famine has been the success achieved by national and international veterinary campaigns, principally against trypanosomiasis and rinderpest, which have removed natural causes of mortality essential to the attainment of adjustment to chronic Sahel instability caused by climatic fluctuation.

The World Bank report already mentioned was concerned with using the so-called Middle Belt of Nigeria to provide pasture on which to raise cattle for the local meat market. The Nigerian Middle Belt to a large extent coincides, in that country, with the Guinea zone of vegetation. This zone, which stretches across Africa from Ethiopia to Senegal appears to some observers to offer the only refuge for the survivors of the Sahel famine. Schemes are now under consideration for an attack on the trypanosomiases of the Guinea zone. One should note that, whether or not it will provide suitable pasture for cattle, it certainly contains most of the major foci of human sleeping sickness north of the equator; that, given sufficient funds, adequate technical staff and the active collaboration of several African governments nobody could doubt that the trypanosomiases, both in people and in cattle, could be overcome. Whether this would be followed, in the long run, by beneficial results is a question about which the present writer is very doubtful. Unless accompanied by what would amount to major revolutions in the political, social and cultural life of these regions, it seems certain that their main consequences would be the expansion of the area liable to catastrophic famine.

In any case the trypanosomiases must be expected to continue to present formidable problems for many years to come. Prior to the

European colonial invasion the inhabitants of Africa had come to terms with the tsetse-fly problem and in the millenia since the appearance of agriculture and husbandry had eliminated tsetses from about 5,000,000 km^2 of the continent. Much training offered by the developed countries has tended to produce white-coated bench workers more ignorant of the problems faced by rural populations than were their own fathers and grandfathers. What is needed is for African governments so to direct training that the rural communities can resume their ancient task of overcoming the infections unavoidable in their encounters with wild-life ecosystems. In this task they must be encouraged to look upon government medical and veterinary officials as public servants over whose operations they must exercise their own share of control. If there is any lesson to be learned from the results obtained by applied science in tropical Africa in the 20th century, it is that vast projects designed to solve artificially isolated problems do not work except with the willing collaboration of the people they are intended to help.

REFERENCE

WORLD BANK 1974 *Nigeria: Options for Long-term Development.* Johns Hopkins University Press, Baltimore and London.

REGIONAL PLANNING AND INTEGRATED DEVELOPMENT IN THE RURAL AFRICAN ENVIRONMENT

A case study of the Mono Region of Dahomey*

MACHIOUDI DISSOU
University of Dahomey

Since the beginning of 1969 an experiment in regional planning has been under way in the Mono province of SW Dahomey. The present study aims to examine the basis of this experiment and its content, and then to proceed to a comparative analysis of the results obtained and of the objectives of development in the rural environment.

I. THE ENVIRONMENT

Situated in the SW of Dahomey, Mono province is a transitional region between Dahomey and Togo. It is a long strip of land, approximately 3800 km^2 in area, lying between parallels 6° 15 and 7° 30 N. and measuring 130 km from north to south. It is bordered by the Atlantic Ocean for 40 km and has a maximum east-west width of 48 km (at Dogbo), lying between the meridians 1° 35 and 2° 10 E.

A) The Physical Environment

1. Geology. The area of Mono province forms part of the geological series of the sedimentary basin of Lower Dahomey which, according to Slansky (1962), is composed of series of mesozoic age (Upper Maestrichian Cretaceous); tertiary (Eocene and Terminal Continental) and quaternary (alluvial and marine series). All these series rest unconformably on the ancient Pre-Cambrian granito-gneissic base.

2. Relief. Mono province is an area of monotonous relief distinguished mainly by a series of plateaux of low altitude and by a hydrological system oriented from north to south, which mark the natural boundaries to the east and to the west of the region respectively. Related to these two principal elements of the relief, one can distinguish three natural zones, which also correspond to the different geological series mentioned above, *viz.*

i) Corresponding to the quaternary series are recent land forms (coastal land forms and the alluvial series of the valleys

*French original.

of the Mono and the Couffo, the two main watercourses of the region).

ii) The tertiary series are represented by:
Terminal Continental in the well-favoured area stocked with oil-palms (*Elaeis guineensis*) which play an important part in the economy of Mono province, as they do throughout Lower Dahomey;
Eocene series in the deposits of Lama, the central west-east depression of Lower Dahomey.

3. *Climate*. Mean monthly temperatures vary between 27° and 31° C. Although variations in temperature are generally small, the northern part (the Aplahoué plateau and the Lonkly area) experiences substantially large variations, from a minimum of 21° C in November to a maximum of 36° C in March.

Precipitation occurs in two rainy seasons alternating with two dry seasons. The extreme NW of Mono province (north of the Aplahoué plateau) is wetter than the south. The average annual rainfall for Aplahoué (over 34 years) is 1102 mm while averages for the coastal zone (over 34 years) are: Grand Popo - 850 mm; Athiémé - 976 mm; Bopa - 872 mm. Further north the Lonkly area has 1607 mm per annum.

In other words, the greater part of the Mono region enjoys a semi-humid tropical climate. Rainfall shows fairly marked regional variations. The north of the province is situated in the southern Sudan zone and the NW is subject to a wetter climate that the south.

4. *Vegetation*. The landscape of Mono, in common with the rest of Lower Dahomey, has been much altered by man. The area is one of derived savanna with a few relics of former deciduous forest cover to be found in sacred groves; Despite a degree of botanical uniformity over wide areas, one can distinguish five main zones:

i) in the north, over the Pre-Cambrian basement complex, is a zone of more or less dry wooded savanna;

ii) in the sedimentary zone, dry savanna disappears gradually, to be replaced by a light wooded savanna or savanna with shrubs (*Daniella Oliveri, Parkia biblogosa*, etc.);

iii) in the wettest parts of the sedimentary zone on the southern plateaux (Athiémé) the vegetation is 'forest bush', an unstratified cover composed of trees 2 to 5 m high with in rare instances a slight ground cover of vegetation. This is the region of subspontaneous oil-palms, varying in density generally according to human population pressure;

iv) low-lying areas, the valleys of the watercourses and areas around coastal lagoons have their own characteristic vegetation (*Dialium guineensis*);

v) finally, the coastal zone, formed by sandy alluvial deposits of the rivers and the sea, has been planted with coconut palms.

5. *Soils*. The differentiation of soils in Mono stems from the combined effects of geology, climate and relief. One can distinguish five types of soil:

Pale brown somewhat ferrous soil, categorised according to topography (that found on slopes, at the bottoms of slopes, and in

low-lying areas);
Calcimorphic soil, occurring over Eocene outcrops;
Somewhat ferrous soil, extending over the greater part of the plateaux of Mono;
Hydromorphic soil, found either on the shores of lakes (lakes Toho and Ahémé) or in lagoon depressions on sandy stretches of the coast;
Recently formed soil over the alluvial series (recent river-born alluvial deposits) of the Mono and Couffo.

B) The Human Environment

1. The Peoples of the Region. Mono province contains about 10 ethnic groups. By far the largest group is the Adja, to which belong the Agassouvi who come from Tado and were the founders of the kingdoms of Allada, Abomey and Porto Novo. There are also several sub-groups of the Adja, including the Sahoué (East-Athiémé); the Adja-We (of Aplahoué); the Watchi (in the frontier region of Athiémé); and the Tchi (Athiémé). There are also: Fon and Kotafon in the Athiémé district; Hwéda, a sub-group of the Kpla, on the shores of Lake Ahémé; Kpla in the Grand Popo district; Mina at Agoué. Aïzo and Yoruba, who do not originate in this region, are found throughout the province, especially in the most important commercial centres.

2. Demographic Data. On the basis of demographic surveys carried out in 1961 by INSEE (the National Institute of Statistics and Economics), Paris, the population of Mono on 1 August 1969 is estimated at 367,000 and on 1 August 1970 at 378,000. This population is essentially rural: none of the main centres (Athiémé, Grand Popo, or even Lokossa, capital of the province) can be classed as urban according to the survey. In the absence of recent censuses in the region, the same criteria of towns and urban population used in the 1961 INSEE survey have been retained for the demographic data.

The average population density has increased from 80.9 per km^2 to 97 in 1969, 99 in 1970. There are, however, significant differences between the various sub-regions. Whereas some areas are very sparsely populated, such as the Lonkly and Tchi plateaux, in others the density rises well over 100 per km^2 (Athiémé, Grand-Popo, Bopa).

The proportion of children in the population of Mono is increasing; and now nearly 50% of the inhabitants are less than 15 years old. As in most of Dahomey, the population as a whole is characterised by an imbalance in the distribution of the sexes: there are 89 men to 100 women, and in the 19 to 59 age group even 77 men to 100 women. This pattern is typical of rural zones where there has been out-migration of young men, who abandon rural areas in the hope of finding employment in the towns. Many of the women are active in agriculture in Mono, where the rate of female participation in this sector (93%) is the highest for all Dahomey. This represents compensation for the economic imbalance resulting from the lack of adult men.

The relatively low rate of schooling - 28% for boys of school age and 8% for girls (as against 30% and 13% respectively on average for the nation as a whole) - is but one indicator of the social underdevelopment of the province.

The birth rate (5.5%) and fertility (0.240) are the highest in Dahomey; the mortality rate (2.6% as opposed to 2.0% for Dahomey as a whole) is also among the highest; the rate of natural increase of the population is thus 3.0% as against a national average of 2.7%.

C) Man and his Environment

1. The Various Factors affecting Human Activity. Human activity in Mono is circumscribed by the environment, the constituent elements of which are: the natural, demographic, socio-religious, socio-historical, socio-political and socio-economic factors. Any external plans for development, whatever their scope, must take these factors into account.

a) Natural and demographic factors. The population of Mono is mainly rural in origin and, as our survey results demonstrate, practises activities linked with the rural environment, notably agriculture. Agriculture in this province has certain characteristics which it is important to underline. In Mono, the people are first and foremost peasants, not commercial farmers, that is to say, agriculture, whether engaged in as a primary or secondary occupation, is vital for the family's subsistence and is therefore a way of life rather than an occupation. Agriculture tends to be a secondary occupation, however, among the inhabitants of the southern plateaux where tertiary sector activities are more important than in many other parts of the region. This is the result of a higher concentration of socio-economic infrastructure investments in the coastal zone compared to elsewhere. In Mono there are, in addition, more women whose main occupation is agriculture than in the other provinces of Dahomey. They represent about 69% of the category of farm labourers - a category which corresponds to 37% of the total number in farming.

The system of exploitation of the land is a peasant system, directed by the head of a polygamous household. An average household consists of 5 adult members, only two of whom are active in farming - which indicates agricultural underemployment on a fairly large scale over the province as a whole. Estimates based on various surveys have shown that the difference between the normal capacity for agricultural work of the population and the amount of work normally done varied between 2,265,000 days on the lowest estimate and 3,460,000 days on the highest, reckoning needs in terms of farming days according to area cultivated and local methods used. Such a high rate of underemployment has a variety of causes, originating mainly in the natural conditions with which agriculture has to contend. The farmer in Mono, as in most rural areas of Dahomey, is limited in his ability to use the land, and particularly, in terms of realising its potential, by his technical resources: in fact his traditional tools allow him little mastery over the heavy clay soil whose potential is in consequence far from adequately exploited.

Furthermore, the inability to master appropriate techniques - water control for the development of the alluvial lands of the Mono and Sazué, drainage of the hydromorphic lands of the river delta and the coastal zone; the control of fire in the grassy savanna, for example - serves to diminish the land resource base of the region. Cultivation practices typical of subsistence agriculture are widespread, notably crops grown over burned bush, insufficient use of manure, intercropping, etc.

b) Socio-religious factors. Religion has an incontestable power over the life and behaviour of the Mono peasant; and agriculture does not escape its influence. This becomes obvious if we examine the factors of production such as the land and work.

Traditionally land is an inalienable good and its successive occupants enjoy only the use of it. It is true that nowadays these fundamental customary laws of ownership, like most African traditions, have changed profoundly through contacts with foreign societies; yet one should note that tradition remains deep-rooted in the rural areas of Mono, for economic, social and political reasons which are discussed below. All development projects must take account of this context, allowing for the special characteristics associated with land, in particular by defining the system of production, the ways of land use and of occupation of land.

Work, another of the factors of production, is, in respect of its importance and frequency, also influenced by religion. For example, the large number of ceremonies necessitated by religion interrupts agricultural work in the fields. Studies we have carried out in Lower Dahomey, whose results are also valid for Mono, indicate the following:

Occupation	Average number of days per year	%
Agricultural work	100	27.39
Rest	148	40.54
Illness	34	9.31
Travel	30	8.21
Funerals (religious ceremonies)	21	5.75
Domestic work	17	4.65
Other	15	4.15

The table thus shows that funerals, which are only one of numerous aspects of religious observation in rural Dahomey, take up 5.75% of a peasant's time, which is in turn equivalent to 21% of the time devoted to agricultural work.

c) Socio-historical factors. The history of the settlement of the different peoples of Mono governs the conditions of their occupation of the land, that is to say the relations existing between the various groups, which in their turn have important implications in the sphere of agriculture. These socio-historical factors, closely linked to the socio-religious ones, have helped to forge the legal framework within which agriculture is carried on in rural areas, by defining the different traditional ways of acquisition of family lands (by inheritance, by gift) and of occupation and exploitation of land (wage-labour, loan, rent).

In the light of these socio-historical factors we can look at the existing relationships between two large ethnic groups of the Bopa plateau in the SE of the Mono region. The Hwéda (a sub-group of the Kpla or Hula), living on the shores of Lake Ahémé, are traditionally fishermen. The Sahoué (a sub-group of the Adja), living on the Bopa plateau and as far as Athiémé, are great farmers and occupy all the agricultural land. During the last few years, undoubtedly as a result of a population 'explosion' and the use of techniques ill-adapted to the circumstances, grave problems of overpopulation have arisen among the fishing peoples of Lake Ahémé. Over and above the difficulties, which are already apparent, of diversifying even partially their source of livelihood, the Hwéda face land problems, since in general on dry land they possess only the small plots on which their huts stand. According to some traditions, the Hwéda, whose place of origin also seems to have been Tado, whence they emigrated to the shores of Lake Ahémé, were forcibly restricted to fishing by the first occupants of the land. This historical condition said to have been imposed on the group must have systematically isolated them from the process of opening up the land of the plateau. Solutions to this problem must arise, therefore, out of this socio-historical context.

d) Socio-political factors. The geographical situation of Mono and the small size of its population have led to the isolation of the region. By analysing this isolation it is possible to ascertain the causes of the cumulative underdevelopment from which this southern province has suffered despite the fact that it contains about a third of Dahomey's coastline. The socio-political factors can be examined, together with the historical aspects, in the light of problems whose analysis is outside the scope of this present paper. (For example, one might wonder why and how the Adja people, ancestors of the principal ethnic groups of Lower Dahomey - Fon, Goun, Aïzo, etc. - have not been able to maintain their socio-cultural and linguistic unity.)

The causes of the isolation from which Mono suffers, however, may properly be sought in contemporary socio-political factors. A rapid review of contemporary political history in Dahomey may clarify the problem. Colonial politics, exacerbated by 'tribalism' and the choice of a statesman from SE Dahomey, a Goun, as the first Dahomean national deputy, triggered off a series of regional and ethnic groupings. On this basis were constituted three large federated groups of traditional political parties supported by electorates whose original motivation and ways of functioning sprang from nothing but ethnic loyalty. In this battle for allegiance, which shaped the ethno-political forces of modern Dahomey, Mono - the home of the Adja people, cradle of the dynasties and peoples of Allada, Abomey and Porto Novo - as if condemned by history for not having achieved the unity of its people, has remained a 'country' without a leader. Since then, it has been a bone of contention between the two main ethno-political bastions of south Dahomey, which have divided the electorate between them. The fact that Mono is, in effect, a 'leaderless' region has in the past led to the division of the population by sharpening their clan loyalties and is one of the fundamental causes of the poverty of the region.

e) Socio-economic factors. One important consequence of the socio-political situation of Mono is its cumulative underdevelopment in socio-economic terms. Mono has long been isolated, ignored, and separated from the main centres of decision making. In these centres intrigues and ethnic interests have inevitably helped to accentuate and maintain the inter-regional imbalance of Dahomey, although this originally derived from various natural circumstances. The absence or inadequacy - in sum the poverty - of socio-economic infrastructure has kept Mono an enclave with few of the lines of communication with the outside world which might otherwise have helped the development of commerce and stimulated agricultural production.

2. *The Traditional System of Land Tenure and its Evolution.* As has already been said, the traditional system of land tenure derives from, and is bounded by, socio-historical and socio-religious factors. In Lower Dahomey most customary law asserts that land is a good which belongs to the ancestors; it cannot be alienated; the living only enjoy the use of it; and it is transmitted in succession to the descendants of the clan or lineage through the mediation of the chief of the clan or lineage, who is the guardian of the land.

Nowadays, as a result of various internal factors (population pressure) and external factors (contact with other areas) this principle has undergone important changes, involving, for example, the partition or alienation of clan or lineage property. The impact of these changes varies from one ethnic group to another according to the modifications which have taken place in traditional social structures. Paradoxically, although Mono is a region where migratory movement has been extremely limited because of poor communications, its traditional social structures were disrupted very early on. The extended polygamous family has disappeared in favour of the restricted polygamous family, and this situation has had profound repercussions on the system of land tenure. Large tracts of land owned by the clan or lineage have given way to smaller territories exploited by families, their size depending on the population density and type of soil (whether plateau or lowland). (For a sketch of the distribution of different crops, listed according to type of exploitation and sub-region of Mono, cf. Dissou, 1970, vol. II.)

3. *The Main Occupations of the People.*

The Primary Sector.

a) Agriculture: The traditional framework of agriculture in Mono remains exploitation of the land under the authority of the head of the family. The workforce consists of the members of the household (about five active members in each, only two of whom are involved in agriculture), and wage-earning agricultural workers (employed manpower). Certain of those in charge of agriculture sometimes have recourse to a system of mutual help in busy periods; but in general the farmers of Mono are individualists and do not favour any form of pooling of land resources for communal work. The principal cultivated crops are: maize, cassava, cotton, tobacco, rice, groundnuts, coffee and yams.

b) Stock rearing: Animal husbandry has a negligible place amongst the peasants' activities, for Mono is not in fact a stock rearing region. It is not an independent activity even in areas where it has relatively importance, notably in the Bopa district; it is simply one of the elements of agriculture. Cattle, sheep and goats are the dominant species on small family farms.

c) Fishing: Fishing is the main activity of the peoples of the shores of Lake Ahémé (Hwéda or Kpla) and those of the sea coast (Kpla or Hula) in the Grand Popo district. Fishing in inland waters (Lake Ahémé) by traditional methods (mostly *acaja* in the eastern part and basket fish traps in the southern part) has run into serious difficulties in the last few years because of over-population of the lakeshore. (See p. 78)

The Secondary Sector.

The craftsman class is almost non-existent in Mono and processing of agricultural produce is confined mainly to the treatment of cassava to make *gari*.

The Tertiary Sector.

Traditional trading activity is unevenly distributed throughout the region. It radiates out following the development of the present road network, and the number of roads is highest in the south of the province because of the coastal axis joining Cotonou to Lomé (Togo). In consequence, on the Comè plateau and in the lowland areas, which have better access to the communications network, most of the inhabitants prefer trade to agriculture. This trade consists mainly in foodstuffs (agricultural produce and basic necessities), construction materials and beverages. Two categories of trader are involved. The marketing of agricultural produce is generally the monopoly of women, who sell to small traders, local agents of export organisations, where the initial collection of raw materials (notably palm products) is concerned. Trade in basic necessities remains the main concern of small traders who keep shops in the villages and in the main commercial centres. Their activities serve to channel into these areas certain imported or locally manufactured products, supplied by shops in urban centres.

Trading takes place in periodic rural markets, integrated into the commercial network of the region. The principal markets are: Comè, Dogbo, Azové, Klouékanmè, Lobogô.

The prices of agricultural products, very sensitive to alterations of supply and demand, vary considerably within the region at different times of the year. The continual fluctuations in price in these micro-markets expose the peasant's income to the risk of serious depreciation, so that administrative measures (the fixing and effective control of purchase price paid to agricultural producers) need to be introduced to protect them sufficiently.

II. EXTERNAL INTERVENTION: PHYSICAL PLANNING IN MONO

The analyses of the first part of this article show clearly that Mono is a predominantly agricultural province. Like all other regions of Dahomey, Mono has been the theatre, the scene of operations in the field of agriculture, in which the different agricultural policies pursued in the country during the various periods of the nation's life have been acted out. We can identify three broad phases of regional and agricultural development policy in Dahomey - the colonial period; the dawn of independence; and present trends (Dissou, 1970). The present study is limited to tracing the main outlines of these 'policies' and examining their various aspects and repercussions for the Mono region.

A) The Colonial Period.

As in most colonies, the agricultural policy of the colonising power pursued certain objectives which, in Dahomey, may be defined as follows: in general, the economy of the colony constituted an appendage of Europe. Thus, the programme of economic development was based on encouraging the growing of 'export' crops. These were introduced in each region after experiments in research stations. In fact an agricultural 'region' at that time equalled the area where greatest efforts were being made to extend the cultivation of a particular export crop.

Among the many consequences of this policy the most notable were the institution of monoculture, which still persists and from which Dahomey, like other former colonies, suffers serious economic consequences; and the abandonment of food crops in favour of export crops, which has had the effect of diminishing the people's self-sufficiency in food. In this context, Mono province - which was part of the second agricultural district (*circonscription*) - was consigned to the development of coffee, cacao and coconut palms.

If these crops prospered to a certain extent in the colonial period, they now have a minimal place in the economy of Mono. In fact most of the trees are now completely untended and yields have diminished considerably. Apart from some fairly ineffective rehabilitation operations undertaken here and there, coffee growing remains secondary everywhere and cocoa growing is now virtually non-existent, despite the enormous potential in this field.

These efforts at development took place within the framework of peasant agriculture. The help given to growers consisted mainly in the distribution of seedlings or cuttings and in technical assistance, however irregular, from a small staff. The inadequacy of resources available to the peasants explains the lack of attention given to these plantations which were built up with improved and recently introduced plant strains requiring much financial investment and sound technical knowledge.

The introduction and encouragement of export crops under these conditions had important consequences, namely, the disenchantment of growers in the face of results which did not measure up to their expectations. This factor has often defeated subsequent programmes for the promotion of these crops. Because the administrative framework was rigid and the crops perennials, agricultural policies applied during the colonial period have had a marked effect on the orientation of subsequent programmes.

B) *Early Independence Period*

After independence the main emphasis of the colonial period was at first maintained, but a certain willingness on the part of the administration to increase and reinforce the resources of local rural organisation began to emerge. The system of administration, however, showed no change in its philosophy. Aid to the peasant masses continued to come from administrative bodies still organised, according to the French model, in the trilogy of Agriculture, Rivers and Forests, and Stockrearing.

Aid from the former colonial power, transformed into bilateral aid, emerged - alongside the national extension services - in the shape of development companies in the various regions. Another characteristic of this period was that the cooperative movement which until then had played only a secondary role in peasant farmer development schemes grew apace thanks to the interest taken in it by the new political leaders, who made it an increasingly important element of their agricultural policy based on regional reorganisation.

This newly-defined policy of regionalisation was henceforth to be incorporated in the new administrative system of Departments created after the institution of the Republic (4 December 1958). Thus, on the organisational level, the former agricultural district (*circonscription*) which sometimes embraced three Departments disappeared in favour of the new 'agricultural region', corresponding to the territorial limits of the administrative Department or Province. The former agricultural region of Mono, under the authority of a Head of the agricultural region, contained sections at the level of the former sub-prefectures - today called districts - responsible for modernisation. The staff, always insufficient in numbers, operated at the level of villages or with groups of peasants, but in an uncoordinated way and with insignificant effects on production. Any increases in production recorded resulted mainly from extension of the area under cultivation, a consequence of population growth.

The establishment of development companies in the context of bilateral aid later led to the setting up of CFDT (The French Company for the Development of Foreign Fibres and Textiles) on the Aplahoué plateau in the north of Mono province. At that time the agricultural region was divided into two sub-regions: the organisation of peasants in the northern sub-region (consisting of the former sub-prefectures of Aplahoué, Klouékanmè and Dogbo) was entrusted to the CFDT which concentrated mainly on the cultivation of cotton. The southern sub-region (consisting of the administrative districts (*circonscriptions*) of Athiémè, Bopa and Grand Popo) was the responsibility of the agricultural service.

This dichotomy in development framework gave rise to the creation, in the heart of the region, of two zones of markedly unequal development. The cotton-growing triangle (Aplahoué-Dogbo-Klouékanmè), in the northern savanna area, enjoyed relative prosperity in comparison with the rest of the province, where the policy was one of extension work centred on the traditional food crops. In fact the two types of development activity (one based on the government administrative structure and the other on the development company), were endowed with different resources, and used different methods - one more diffuse, the other more concentrated.

In addition, CAITA (the Agricultural and Industrial African Tobacco Company) sent its staff indiscriminately into both sub-regions to encourage tobacco-growing among the farmers. Furthermore, in the Mono province, SONADER (National Rural Development Company) supervised another experiment within the framework of COOPAHA which involved an industrial plantation of selected palm trees (4,015 ha.) planted on the edge of an area of rural development.

This series of disparate, sometimes uncontrolled and uncoordinated actions, inspired by different methods and concepts and using varying resources, rapidly created discrepancies and gaps from which the peasant was the main sufferer. The process of development at the level of the region (which is the same as the department or province), as observed in Mono, was almost identical with and took the same form as in other regions of Dahomey. In general, this kind of regional development showed the following characteristics: the various development projects of organisation administered in different ways lacked coordination, despite the fact that the principle of unity of command was followed through in the appointment of th

Perfect as the one with overall responsibility for the programme of regional development. (This was a direct consequence of the coincidence of the agricultural region and the territorial limits of the former department or present province.)

A study of regional development in Mono in the early phases brings out following points, therefore:

1. *Production.* The multiplicity of extension organisations concerned with the same peasants, under the conditions already discussed, had disastrous consequences which needed to be corrected: the peasants themselves were completely confused by the presence of officials answerable to different bodies, in some cases giving contradictory advice, and each mainly concerned to 'put over' his own crop. The staff in their turn, working with unequal facilities and belonging to organisations having differing philosophies, became lazy on the one hand, on the other dynamic. In evaluating results difficulties arose even at the sub-regional level, because it was not always easy to demarcate the spheres of activity of the different development bodies. The same problems occurred in drawing up reports. The concentration of considerable technical and financial resources on the 'reserve' of 4,000 ha. of selected palm trees and its oil mill, whose capacity of 25,000 tonnes of palm fruit per annum, incidentally, was never achieved, helped to make COOPAHA an 'island of illusory prosperity', whose problems could only be resolved within the context of regional development which would allow some degree of harmonisation within the whole economy.

2. *Marketing.* Experience has shown that agricultural productivity increases rapidly where the marketing is properly organised, especially with respect to the initial collection of the produce and price stabilisation. Cotton growing is a case in point: the very recent introduction of the *Allen* variety (more productive than the *Barbadense*) has very quickly won favour with Mono peasants.

The lack of roads compounds the problems of marketing in Mono, as it does the problems of supplying the local population with basic necessities. Also, in Mono more than elsewhere, because of the almost total absence of rationalisation of trade and the commercial network, the peasants, doubly exploited by hordes of petty traders and middlemen, have little incentive to increase agricultural production.

3. *Financial aid to farmers.* Organisation up to the time in question did not allow of any financial aid to the rural areas; even though it was realised that such aid was vital to support any extension work aimed at increasing agricultural production. It is true that CFDT had introduced a built-in means of supplying and charging for certain aids to production (such as insecticides and fertilisers), but the producer played no part in the decision making in this process, so it can hardly be called agricultural credit in the true sense of the word.

At a higher level the Dahomey Development Bank (BDD), operating within the conventional framework of financial institutions, tried to operate a system of agricultural credit completely unadapted to the circumstances. In fact the rates of interest charged were far too high; in the absence of a cooperative organisation, which would have allowed effective guarantees on loans through collective responsibility, the conditions imposed were exorbitant.

In the light of all these considerations, the CARDER (Regional Action Centre for Rural Development) plan was formulated, to combat the inadequacie of the previous system which we have been describing. The Mono CARDER came into existence officially on 1 January 1969.

C. Present Trends: Objectives and Organisation of CARDER[1]

The objectives of CARDER derive from the context already discussed, that is to say, the technical coordination of all agricultural activities was henceforth to be entrusted at the provincial level to a Head of Operations (generally an agricultural engineer), the Director of CARDER. CARDER is a public commercial and industrial organisation with corporate civil identity and financial autonomy. It must perform certain functions, notably, it must promote:

i) An increase in agricultural production, by the diffusion of technical knowledge among individual farmers or groups of producers in order to bring into general use improved methods of cultivation;

ii) The rational organisation of the marketing of all agricultural produce (both export and foodstuffs) by helping to form village groups of producers, by proper planning of the commercial network, in order to reduce transport costs to the minimum, so that producers can be supplied with the means of production (fertilisers, pesticides, etc. and basic necessities);

iii) Adequate organisation of agricultural credit;

iv) Progress towards industrialisation by encouraging the exploitation of agricultural produce.

The organisation of CARDER focuses around two bodies: the central executive (*Direction Central*), led by the Head of Operations (the Director of CARDER), and the administrative council. The latter is the decision-making body, presided over by the Prefect. Most of its members are the Directors of services provided by the Ministries of Rural Development and Economy, others are Heads of the administrative districts of the province and others representatives of the provincial groups of producers.

The *results* of CARDER's activities - its effectiveness during the first six years of its existence - may be assessed under three headings: administrative, technical and financial.

1) Administration. During the first years of its existence, CARDER in Mono has functioned mainly under the tutelage of SONADER. Its staff were provided by the Agricultural Service and guaranteed by the Public Service Statute. CARDER's own statutes have not been sanctioned by the govern-

[1]It is useful to note that a system of regional development analagous to CARDER has existed in Atacora Province since 1966 under the name of the Atacora Operation.

ment, so that it cannot directly assert its autonomy. The civil service origins of the staff have been a far from negligible handicap to the Directors as they have considerable difficulty in applying sanctions to defaulting officials. The double tutelage (of SONADER and the Agricultural Service, which continued to exercise some ill-defined legal responsibility) has also been a restriction of considerable importance.

2) *Technical programmes.* The goals - both qualitatively and quantitatively - are defined by CARDER in agreement with SONADER. Programmes covering all aspects of CARDER's activities are drawn up at the beginning of each agricultural season after the evaluation of the results of the season before. The activities of CARDER in Mono have covered several fields, the most interesting of which are: agricultural production, marketing and community development.

a) The agricultural programme consists in promoting cultivation techniques in each sector (sub-region) by means of a pilot cultivation plot determined according to the agricultural potential of the sub-region. The extension work is usually done among groups of farmers, very rarely individuals. The farmers' fields are grouped either into blocks of cultivated land, in the case of fairly consolidated groups of farmers, where the work can be done collectively, or in contiguous individual plots, in order to make the extension training most efficient.

The principal crops for each sector are:

SECTOR:	PILOT CROPS:	OTHER CROPS:
Aplahoué	Cotton	Maize - Groundnuts - Tobacco
Dogbo	Cotton	Maize - Rice - Groundnuts
Klouékanmè	Cotton	Maize - Tobacco - Groundnuts
Lokossa	Maize	Rice - Cotton
Bopa	Cassava	Maize
Grand Popo	Maize	Rice

Objectives in terms of area under cultivation and yield are defined according to crop and sector; all these elements are determined in the light of the degree of assimilation of techniques by the farmers. One may note, however, that degrees of intensification are not rigorously defined. As far as results are concerned within the province, the Aplahoué-Dogbo-Klouékanmè triangle which corresponds to the extent of the cotton-growing zone seems to show a certain amount of progress in a number of fields.

b) Marketing. Those responsible for CARDER quickly saw that agricultural production could not be stimulated in a region such as Mono, which had so long been isolated, without first attacking the problems of marketing agricultural produce. For this reason, one of the first tasks to which CARDER applied itself was the organisation of the marketing of paddy rice and cotton. Indeed, the extension of the area under rice cultivation quickly raised difficulties of milling and disposal; this thorny problem had to be solved and the new organisation has indeed been able to ensure the marketing of 50 tonnes since its inception during the 1969-70 agricultural season.

The initial collection of cotton which had normally been done by the CFDT was taken over by CARDER as soon as it was set up. Later, the ginnery at Hagoumè arranged the processing of the cotton produced in Mono. The aid provided by CARDER to farmers in the field of commercial cotton-growing comprised mainly: the improvement of productivity (technical advice; the supply of chemical fertiliser); the organisation of marketing (grading and purchase of produce); processing the product (ginning); preparation for export .

During the last two years CARDER has organised the marketing of farmers food crops, notably maize, thanks to the facilities for pre-drying and storage offered by the Hagoumè silos, whose capacity is 2,000 t. of grain. In general, the organisation of agricultural marketing has helped to break through major bottlenecks, to improve the monetary income of the peasant masses and to reduce the involvement of middlemen especially in the trade in raw materials, such as palm products - trade which is now organised by village groups created with the help of CARDER staff.

c) Community development and extension work carried out by CARDER is based on groups of producers in one village or a group of villages, set up with the technical assistance of the Cooperative Action Service. These village groups are the framework in which CARDER conducts its whole programm production, marketing, literacy, health and nutritional education.

An ever-growing number of village groups is being established in the various sectors of the province. Although they are active in many fields, their main concern is in the field of marketing (the initial collection of the produce and supply of basic necessities to the members of the group). The rapid increase in the turnover of these organisations and of the returns to their members is eloquent testimony to their dynamism.

In the cotton zone, the literacy programme plays a relatively important role. Thanks to this particular drive for literacy most of the groups are involved in all the operations of the marketing of cotton, which brings them considerable advantages. Among the women's groups in villages, health and nutritional education, provided by community development workers, has been of much benefit to the women.

3) Finance. One of the main advantage for CARDER of the tutelage of SONADER is the financial aid it receives. This situation is due to the fact that CARDER does not yet have independent status and, less importantly, due to its lack of credit-worthiness with the banks. The financial aid from SONADER has an important implication, however, namely, that all the financial transactions of CARDER in Mono are effected through SONADER which holds a separa account for CARDER.

D) Inadequacies and Omissions of CARDER in Mono - an Inventory of the Unresolved Problems of Integrated Rural Development.

An examination of the quality of the results obtained by CARDER during the first six years of its existence shows the following failures and omissions first in regard to its aims, secondly in regard to its methods.

Aims. Because of CARDER's concepts and its system of organisation, the staff assigned to Mono does not aim in the short term or even in the middle term to coordinate the whole of rural development. In its listed aims CARDER's activities cover only the strictly agricultural field. Even stockrearing, and fishing (an activity of equal importance in Mono) are not directly the concern of CARDER. Apart from this, other aspects of rural development (public health, socio-cultural structures, and economic infrastructure) which certainly need to be taken into account when planning integrated regional development are left out, for the moment at least, of CARDER's brief.

Similarly, we have already mentioned that the COOPAHA, another organisation also under the wing of SONADER, is outside the scope of CARDER. The difficult problems faced by COOPAHA in the last few years could be solved within the context of a programme of regional integration. It is in this direction that lie the immediate solutions, at least, to the situation created by the cooperative's over-manning. This manpower could henceforth be assured of full employment if fields of food crops were planted on the periphery of the cooperative's palm plantation on the 'territory' up to now controlled by CARDER. It should be said that the crisis of under-employment has been aggravated lately by the effects of cumulative drought building up since 1971; the industrial plantation of 4,000 ha. of selected oil palms established in this marginal palm-cultivating zone, has suffered greatly from the terrible effects of this drought. None of the solutions suggested, such as the supply of water to save the palm plantation (which in its turn supplies its local oil mill with 30,000 t. of palm fruit per annum), is economically viable.

In fact, a lake created to the south of this agro-industrial complex could only supply, according to recent estimates, the equivalent of additional rainfall of 500mm over 800 ha. (the approximate area of the lake), which would correspond to a theoretical lowering of the level of the lake by 50 cm. This operation, although technically possible, taking into account edaphic and topographical conditions which would permit gravity irrigation, would nevertheless, not be economically viable. Another solution would consist in drilling boreholes at intervals around the perimeter. Like the first solution, it must be rejected for economic reasons. (It should be mentioned, however, that an attempt at irrigation through gravity over a small area inside the block has shown a large increase in yield from 2.5 t. per ha. to 14 T. per ha.).

To avoid the continuing diminution of the income of the members of the cooperative a programme of supplementary activities has just been devised. The great difficulties which have beset the COOPAHA in recent years demonstrate, however, the clear necessity of turning immediately towards integration of the different projects undertaken in the province of Mono. We have mentioned a supplementary programme aimed at this objective; the same considerations will apply when it is necessary to widen the area of collection of palm produce to ensure better supplies to the oil mill, as the raw materials (palm fruit) coming from the industrial plantation are proving inadequate. (Yields now vary around 3 to 5 t. of palm fruit per ha. against a forecast of 8 t. per ha.) It must be admitted that this solution presents technical problems because this modern oil mill can only treat limited quantities of the produce of the surrounding wild palm trees.

Methods. When discussing CARDER's methods of approach and extension work programme for bringing new techniques to rural areas, it seems vital to reexamine present thinking on the subject in the light of the results obtained and the goals set. Naturally all methods of approach and execution are planned with a view to achieving rural development. To reach this goal of development necessarily implies quantitative changes and qualitative improvements. In addition, all rural development must aim to increase agricultural production (by increasing productivity); it must also at the same time effect socio-economic changes, that is to say, social modifications and improvements in income.

Project evaluation is often confined to the question of financial success; thus the failure or success of a programme is judged on purely quantitative evidence, that is to say, according to production graphs. Similarly, the efficiency of administrative staff has been evaluated hitherto in terms of acreage under cultivation and production figures. This method of judging results is the consequence of the choice of experiments - which, as often as not, concern export crops (mainly cotton and palm products in Mono), carried out as part of programmes financed by foreign aid. The inadequacy of such programmes is now well enough known, namely, the progressive diminution in foodstuffs production, exacerbated by population growth, the result of which is acute food shortage.

It is necessary therefore to reexamine concepts concerning development in the light of such inadequacies. Rural development must necessarily aim at promoting both quantitative and qualitative change - that is, modification of the existing socio-economic structure. In this context, all administrative and extension work must aim to help farmers to run their own affairs. In fact, the best form of aid is one which aims to eliminate itself as soon as possible.

REFERENCES

DISSOU, M. 1970 *Régionalisation et Développement Agricole au Dahomey*. Thèse du 3e cycle, EPHE (Ecole Publique des Hautes Etudes), Sorbonne, Paris.

SLANSKY, M. 1962 'Contribution à l'étude géologique du bassin sédimentaire cotier du Dahomey et du Togo' in: *Mémoires du Bureau de Recherches Géologiques et Minières*, no. 11.

THE ETHNOLINGUISTIC APPROACH TO THE RURAL ENVIRONMENT: ITS USEFULNESS IN RURAL PLANNING IN AFRICA*

PHILIP LANGLEY
Environment Training Programme, IDEP-UNEP-SIDA
Dakar

However little effort he makes to listen to the social groups for whom he is supposed to be working, the adviser will soon wake up to the fact that there is a certain discrepancy between his own ideas and theirs.[1]

In his work, whether as economist or veterinary surgeon, agronomist, planner or doctor, difficulties arise in the carrying through of projects however carefully put together and calculated on the basis of enquiries, research and cost-benefit analyses: these difficulties ought to show him that it is not a case of a mere discrepancy of ideas but, in many cases, a complete breakdown in communication.

The usual interpretation of this type of situation is made in terms of a general lack of widespread education, and resistance to change on the part of peoples clinging to tradition, unreceptive to modern ideas. In some cases it is even blamed on laziness and indolence. In the face of these 'obstacles to development', the remedies suggested are 'changing their outlook', schooling and 'modernisation' by integration into a system of credit and a monetary economy.

> 'The dilemma of the agricultural extension agent originates in a situation where, despite the deep-rooted fatalism, absence of either ambition or enthusiasm, and other traits supposedly characteristic of the peasant's lack of response to new ideas, he is nevertheless finally brought to accept the idea of chemical fertilisers, but then no fertilisers are available' (Abasiekong, 1974).

Nevertheless, this rather technocratic attitude is not the only one possible, and other theories on the way this rift has developed - and hence an appropriate strategy - would emerge from even a brief analysis of, on the

*French original

[1]This paper is not an account of new research in the field. It presents a proposal for one method of approach to environmental planning. This proposal is the result of a collective evaluation of the activities of the Environmental Training Programme, one of whose major objectives is encouraging technicians and researchers to listen to the farmers. We thank Safall Fall, M'Baye Thioune Wade and Jacques Bugnicourt for their encouragement and their friendly criticism in the course of this work.

one hand, the involvement of experts and decision-makers in the social structure of their countries; and on the other, the various methods of elaborating and transmitting knowledge among the different sections of the population, notably at the level of the masses in the towns and rural areas.

Among these hypotheses, one might consider the existence of a popular culture, creativity and even *people's science* (Richards, 1975): their existence has long been recognised in the fields of construction techniques and spatial organisation of the habitat, but is less often admitted in the sphere of concepts, the explanation of social or natural phenomena and economic reasoning. Such concepts, explanations and reasoning, however, are commonplace in the daily life and customs of pastoralists, agriculturalists and the urban masses, and help them to structure their understanding of their environment and interaction with it. Daniel Biebuyck has clearly explained the existence of such concepts and has long underlined the necessity for understanding them:

> 'The majority of African societies have formulated a series of specific ideas and concepts about the land... The difficulty therefore lies in interpreting and giving the proper value to unique concepts which refer to distinct situations and realities... Key formulae, such as 'chief of land' or 'master of the soil'... or categorical distinctions between rights of ownership, possession, use and usufruct... have obscured many analyses and impeded the full understanding of the systems.' (Biebuyck, 1963).

On the other hand, the categories into which we split up concepts which we use as intellectuals or technical experts - formed in the case of most of us in European or American schools - show our own original discipline. These disciplines developed at the same time as, and to a great extent because of, the needs of the economies which dominate the world today. In so far as the use of these concepts and classifications is outside the powers of ordinary people in Africa or elsewhere - and all the evidence points that way - their use can contribute to the spread of a particular ideology and contribute towards reinforcing the domination of these social groups.

Paul Bohannan wrote a critique of the ethnocentric use of western concepts as an obstacle to transcultural studies (Bohannan, 1963). But today, should one not go farther and analyse ethnocentrism and the use of foreign concepts as some of the ideological tools of domination?

> 'Exclusive recourse to mimetic transfer of knowledge, in the form of finite intellectual products and technological processes, is in fact a fallacious short cut to modernity (and) it should be realised that contemporary science and technology is Euro-centric'(Sachs, 1971, p.85).

> 'It is by no means caricature to describe modern science as a European invention which ensures ... military, economic and cultural domination over the rest of the world ... It would not be unjust to say that most researchers and technical advisers become, inadvertently, accomplices in this, fundamentally racist, use of science'(Calder cited in Sachs, 1971, p. 84).

This domination can be domination by socio-economic groups foreign to the society in question, or indeed by groups within the society. It leads us to ask ourselves whether a people's mastery of the environment does not come from the ability to express its true self. (On the revival of a new national movement of self-expression, see Decornay, 1975).

We can approach the hypothesis that a popular science exists by taking as a starting point the people's own conceptualisation of the environment, as found in their languages (Maamouri, 1972). We can observe - within the framework of this article, from the study of rural areas in Africa[1] - a conceptualisation proper to each eco-culture which makes no dichotomy between social and physical (or natural) environment. There will therefore be an 'ecological' character to concepts in many cultures whose dominant mode of production has not led ideology to split into distinct disciplines - one concerned with human groups, the other, with the physical environment. These terms for ecological notions must be taken as cultural or anthropological ideas because 'long before ecology was heard of as a science, people knew the vegetation types of their native countries by simple names which, naive and unprecise as they are, have the advantage of conveying to all members of a human group a definite meaning familiar to them through their common experience'. (De Schlippe, 1956).

On the other hand, the approach of many scientists and technical advisers advocating 'the rational use of natural resources' (Unesco, 1971) tends to limit the environment to physical elements and structures, excluding all consideration of social, economic and political structures (and - more important - excluding the interdependence of these structures and the reasons for their modification). Is this the only possible basis for the study of the environment? Why could not the conceptualisations of rural societies themselves be the foundation of knowledge - which would provide a valid base for formulating action for environmental planning? The question should be asked, at least, if one's aim is that the people themselves should have control over the changes in their environment.

Spatial Concepts and Language

Where study of social categories is concerned, the use of concepts taken from the local language goes without saying:

> 'Let us once again remember the difficulties which confronted European sociology when dealing with completely different cultures, beginning with a vocabulary which sometimes baffled them... The Church-State, spiritual-temporal, ecclesiastic-lay dichotomies where unknown in Arab society in the Middle Ages'. (Sachs, 1971, p. 74).

The use of local concepts is not at present normal practice in the study of spatial organisation.

[1] We have limited our examples to inter-tropical Africa. The same themes that we deal with here could have been tackled with references to other zones, for examples, North Africa (see Bardin, 1965; Bordieu, 1970; Pettonet,1972; Boughali, 1974; Bouzoubaa, et al. 1974).

In every language, however, spatial categories are precisely defined, the very structure of the language being partly a product of the natural space inhabited by the social group and the use made of it. It is often hard to translate directly by a single word in English, French or Portuguese the spatial descriptors in African languages (it should be noted that the opposite is equally true), the result is a set of spatial descriptors with meanings rather different from those of the original language. For example, the words: kitchen, living-room, bedroom, bathroom, house, etc....used to identify spaces belonging to a dwelling in Africa are in most cases as ill-adapted and absurd as the Japanese words would be: *engawa, doma, suljaba, honza, tokonoma,* etc.... (Beardsley, Hall and Ward, 1959).

In examining several sets of categories of spatial descriptors, the first point to note is that the meaning proper to each culture - even each social group - seems to be twofold: inasmuch as it refers to a precise space whose formal organisation is collectively known, and also refers to the habitual social activity in the space in question or to a series of social relations closely linked to this space, including, in many cases, the way the space is used (Biebuyck, 1963).

Each language, then, can describe a whole range of different spatial concepts, and the configuration of space is very much connected with the social structure and the organisation of production and consumption. For the Baoulé of the Ivory Coast, *aoulo* is the name given to the more or less enclosed space, bounded by several buildings, which is the dwelling of a small group of relatives and other dependants of the head of the *aoulo* (Sirven, 1970). In the Fula language, the *gallé* is the family unit which is spatially manifested by an enclosure variable form and nature, enclosed by a mud wall or hedge, among the Tukolor, or, among the Fula of Senegal, by hurdles of interlaced branches, although sometimes it has no fencing but exists as a fairly wide circle of huts and a zone for penning the small livestock. This type of enclosure is one of the main elements of the geographical whole which is the basis for daily life. Although *gallé* denotes the physical space, it also signifies a branch of a lineage, often with dependants, composed of several homes (*poyre*, pl. *poggi*), each forming a separate unit of consumption. A *gallé* constituted in this way may include 3, even 4, generations (Boutillier et al. 1962).

For the Gourmantché of Upper Volta, the *diequ* is the place inhabited by an extended family: the head of the family, his wives and their children, his younger brothers and sometimes paternal cousins and indeed a number of others (a sister's son, a servant, etc.). It should be noted too that, 'like the word *diequ, dansanu* designates simultaneously the unit of accommodation in the physical sense and the group of people who live there. This group corresponds to the nuclear family: the head of the family, his wives, his children, sometimes his younger brothers or his widowed mother' (Rémy, 1967).

Similarly, for the Lela (also of Upper Volta), the space - social relations concept differentiates two levels of dwelling place. The *gui* is a small unit of from one to four rooms side by side. The word also means the group which inhabits them, that is to say: 'a monogamous household'; or 'a household consisting of a polygamous man, his first wife and her children'; or 'a household of a married woman, without the rank of first wife, and her children; or 'of an aged widow' (Barral, 1968). The wider concept is the *kélé*, the family farm, composed of several *gui* as well as other elements, namely: kitchen, showers, barns, poultry houses, the entrance, the yard, etc... Although *kélé* means the place inhabited by the extended family, for a Lela, the meaning is more complex. At the same time as meaning the internal structure of the extended family and the power of the head of the family, it includes spatial arrangements such as the building of barns opposite their 'owner's' *gui*, the situation of the *gui* of the head of the family (the *kélé kebal*) facing the entrance, etc....

Similarly, for the Diola of Lower Casamance, the *hank* is a unit composed of a collective yard around which are the living quarters of the households and the kitchen and nursery gardens, inhabited by several collateral households, and here the herds are gathered in each evening.

In all these examples, neither 'square' nor 'concession' (implying the acquisition of a small area of land from the colonial administration) and still less 'house' is an exact translation of these various words. Other translations are sometimes used, such as 'plot', 'dwelling place', 'enclosure', or 'farm' etc...but none of them render the complex meaning of the original word which can only be translated by a paraphrase.

Another point is the existence for many spatial descriptors of a variable scale. The Hausa word *gàrii*, usually translated as 'town' as well as 'sky', has, in fact, a much wider meaning which is:

> 'human settlement or inhabited place, especially township or village-area (also country or chiefdom), community, inhabitants (of a settlement); (local) economy, including farmland, crops, weather and (local) sky' (Dalby, 1964)

In Wolof, the word *kër*[1], normally translated 'house', refers to the space where the extended family lives and thence implies a series of relationships within the family which are peculiar to the Wolof people. *Kër* also means, on a smaller scale, the nuclear family's dwelling place; and, on a larger scale, 'hamlet' or, in any case, a unit grouping several separate families descending from the first settler in that place. In this case, in fact, the spatial unit corresponds to a lineage segment.

In some cases, without taking on the meaning of 'village', the word '*kër*' is the prefix of a place name: this is often the case when a hamlet or village has developed from one family and its dwelling (its *kër*). The village of Keur Madaro Niang, for example, (6 Km from Thiès in Senegal) has as its kernel the homestead of the present village chief named Niang, grandson of the founder. In this type of village, family links are generally very close and where several families inhabit adjoining plots and descend from the same ancestor, they defer to the eldest, as *borom kër* (lineage head and, by extension, head of the group of homesteads). *Kër* therefore is used on several levels and its meaning takes account of the dynamics of change in socio-spatial organisation.

[1]According to the new transcription of Senegalese national languages (decree no. 71-566 of 21 May 1971). Formerly rendered *keur*.

A third point is the practice - in certain African languages - of grouping categories in broad classes - on one side, 'natural' space, on the other, 'transformed' space. Among the Bassari (of East Senegal), however, certain descriptors are defined in relation to both categories: *andyana* is a part of the land, cultivated by the inhabitants of several plots, which corresponds exactly to a small river basin, the boundaries of the *andyana* following more or less exactly the line of the watershed (Albenque, 1965).

On the other hand, the Diola-Banjal of Lower Casamance (Snyder, forthcoming) distinguish, among others:

a) categories proper to physical geography (i.e. sky; zones which are wooded or covered by other vegetation; lands which can be used for cultivating rice; water);
b) categories of social and human space, that is to say categories concerned with the use of space.

In the first case, it should be noted that these categories are related, in one way or another, to worldwide characteristics of the productive system and its relation to the natural environment: rice growing in mangrove swamps and other places where water can be brought; hunting and gathering in the savanna, woods and forests; fishing in the swamps. In the second case, the descriptors often contain a reference to the act necessary for the transformation (burning, cutting, ditch-digging, fence-building...).

In southern Zambia, among the Tonga people inhabiting the middle stretch of the Zambesi river (Scudder, 1962), 'natural' categories (i.e. the areas where human activities appear non-existent) and also the spatial categories which refer to modifications and human use often imply several parameters in their definition. These parameters are, in different cases:

- the nature of the soil,
- the associated vegetation or typical plants,
- the effect of topography or geomorphology on the agricultural use possible for the social group, taking into account its technical knowledge,
- agricultural potential,
- transforming action (cf. *supra* for the Diola-Banjal).

With irrigation from the Kariba dam, at the beginning of the 1960s, these people (numbering some 57,000) were settled in a nearby area, which, however, showed different natural characteristics and in addition, obviously, there were the lake formed by the dam, and the changes in vegetation, fauna and micro-climate caused by the lake. Because of the social changes which have inevitably followed this displacement it would help to increase our knowledge of the birth of ecological concepts in an eco-culture, to observe what changes have taken place in the Tonga language: modification of categories, changes in meaning or structure between categories, even the addition of new terms. It would perhaps also clarify the ways in which notions of environment and space are passed on: after all, if a science exists, so must a system of education. The same approach would be valid in other large projects leading to population resettlement (Gezira in the Sudan, the Volta river in Ghana, the Niger valley, etc...) and would certainly provide food for thought for planners and the protagonists of the 'sedentarisation of nomads'.

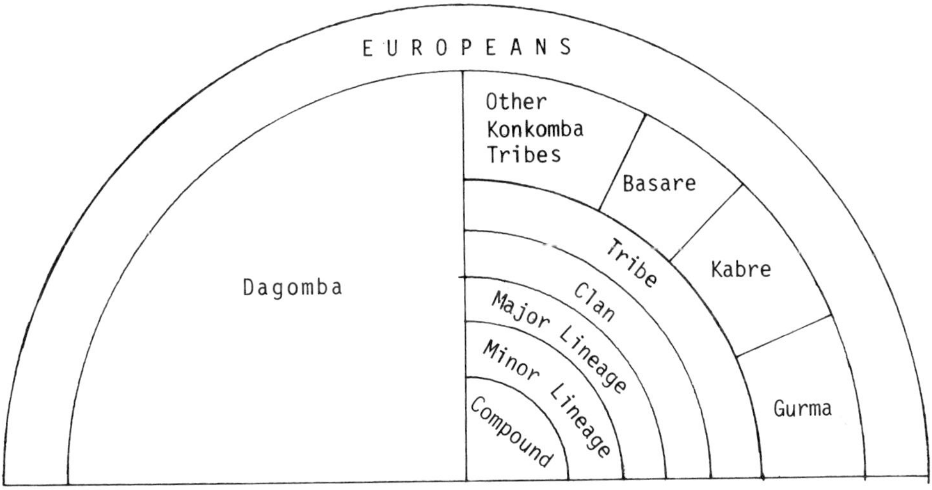

Fig. 1 Structural space of the Konkomba. After Tait, 1961.

Fig. 2: SPATIAL CONCEPTS OF THE NYANGA (KIVU)

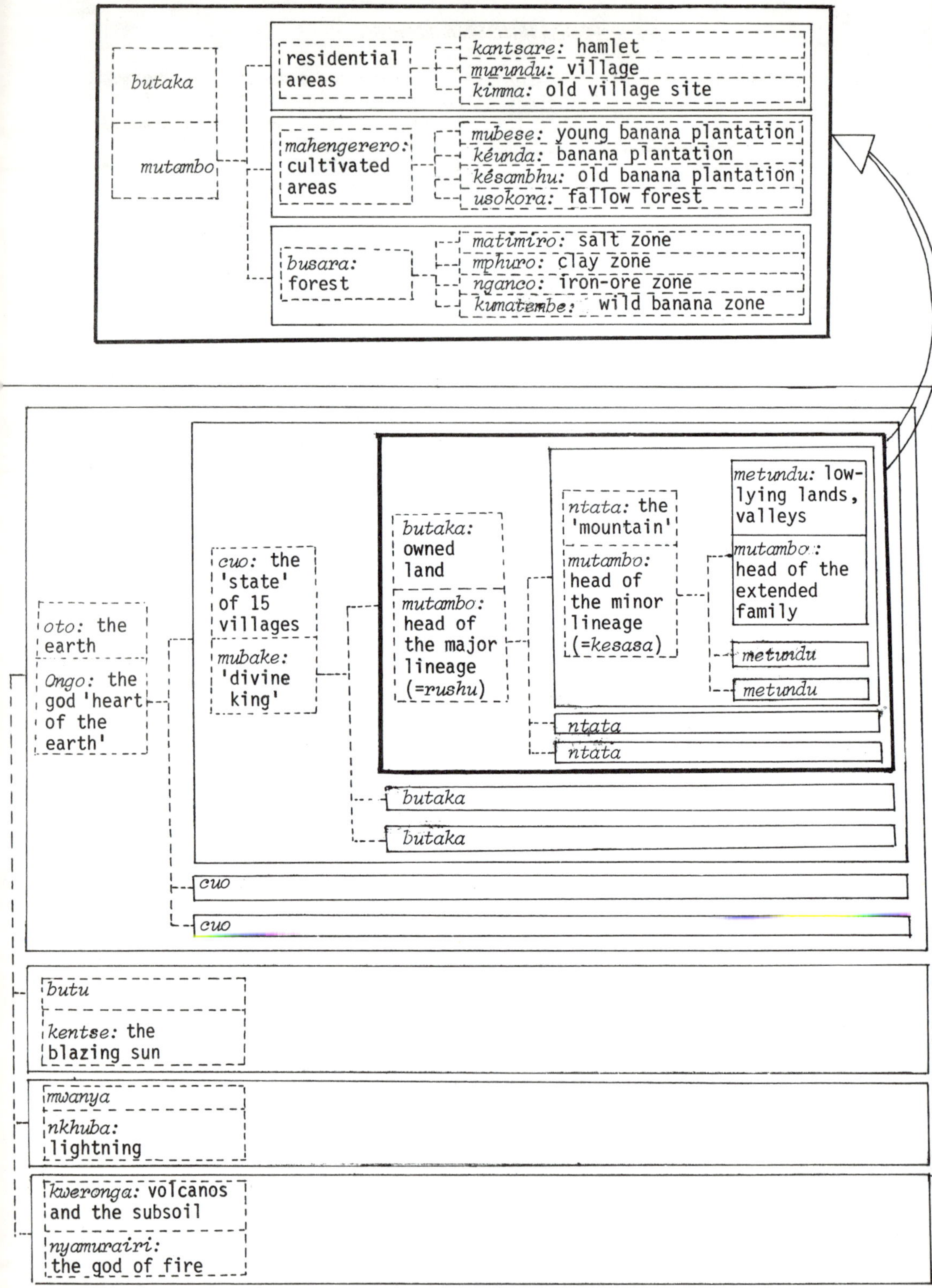

Langley: Adapted from Biebuyck

This multiplicity of parameters for defining socio-spatial categories will oblige us to look more closely at how concepts inter-connect to form a coherent whole. If one wished to speak in terms of a science, one would go further and examine the postulates, theories and epistemology of this whole. These links are, in fact, very fluid among the Konkomba who live in NE Ghana and NW Togo; it is possible to speak of the organisation of an ecological space by referring to the notion of distance between the different elements (see fig.1). Each is a social and cultural unit with values common to all its members and is defined by its social distance from the other elements. Each unit is precisely located physically, but the physical distance does not necessarily agree with the social distance (Tait, 1961).

It has been suggested, from the examples of the Tiv in Central Nigeria and the Plateau Tonga of Zambia that shifting cultivation has been a determining factor in the structuring of a series of relations between social groups, places and spirits, whereby spatial units are not necessarily linked permanently to precise geographical areas, but are defined rather by their relationship to one another. 'As regards land , the "man-to-man" connection is based on agnatic relationship; the "man-to-thing" connection joins man not to a "plot of land" but to a temporary "farm".' These concepts bear more resemblance to topology than geometry (Bohannan, 1963).

On the other hand, the Kikuyu of Kenya seem to have precise and permanent concepts of territorial boundaries giving rise to a 'map' of political units, defined by the geomorphology of the ridges; but having no relation to the exploitation of the environment, which is explained rather by reference to the 'map' of family domains and lineages (Bohannan, 1963).

The close relationships of these structures to cosmogony has been demonstrated for the Nyanga of Kivu, Zaïre (Biebuyck, 1963). Each spatial unit has a privileged association with a person (or god). One says that such and such a person 'is with' such and such a place. The structure is hierarchical, but the parameters defining the structural groups vary according to the different grades of hierarchy (see fig. 2).

All these different examples are ecological concepts par excellence. If the words are hard to classify into the field of 'geography' or sociology', this is because their meaning applies at once to a place, to the social group which occupied the place, to that group's internal structure, to the relationship between the group and the place, and to the way in which the place has been moulded by the group.

The concepts given as examples show great diversity and inventiveness with regard to ways of exploiting the environment, the determining element in the way that the categories are formed being probably the particular relationship in each eco-culture between the system of production-consumption and the natural environment.

Conclusion

On the basis of these facts, our hypothesis is that a social group's perception of its environment - such as it may appear through semantic study - can become one of the main tools of understanding the relationships between social structures and the organisation of space. At the same time, is it not possible that it might form one of the bases for planning to transform the environment by means of the people who actually live in it. Might not the use of the concepts held by the people themselves give them back a certain power in the sphere of analysing the present situation, the identification of problems, the search for possible changes and the formulation of projects: procedures which are often imposed on the people without their opinion being asked?[1]

Let us take as a concrete example, as we did before, the field of housing. A collective analysis of the actual situation of a social group in a rural setting will probably lead to the observation of factors which 'annoy' or 'worry' people: loss of children at an early age; lack of strength at a time of agricultural work; bad health of different members of the family, etc. Research into the causes of this situation will reveal multiple interconnected factors, of which a certain number concern, directly or indirectly, the utilisation of living space and buildings. To eliminate these problems, the whole apparatus of a tiled kitchen with pots boiling on an electric cooker, hot and cold running water in the sink, all adjacent to a WC and similarly tiled bathroom in the same building, seems to be the ideal aimed at, for the near or distant future, by many 'experts' and, who knows, by part of the population as well. If this solution is not adopted in rural projects, it is for reasons of cost rather than rejection of the model. (This is in no way a caricature: one need only look at: a) the position of the utility rooms and b) the way they are labelled in the plans of almost all 'cheap' rural dwellings.)

But might there not be another approach springing from the way of life in, and concentrating on changes and improvements to,existing buildings? Could one not, for instance, put water for washing one's hands near the place where meals are prepared, or put drinking water in receptacles which would prevent mugs being dipped into it or rubbish falling in, etc..? Such changes, while retaining the existing spatial structure, would modify it but allow the assimilation and integration of changes.

It seems to us that such alternatives could be discovered and invented by the people themselves with the assistance of technical'experts', once these latter show their capacity to understand the various concepts involved. We should also sketch rapidly, in a more general context, the different occasions for using categories of space and environmental concepts in cooperative efforts at planning between 'experts' and local people:

a) Analysis of the present situation, including identification, in the people's own terms, of their difficulties and problems: this is parti-

[1]Apart from concepts of spatial delimitation and boundaries, one could extend our proposition to the categories and concepts of those activities (so dear to architects) of the physical and corporal use of space;elements of the socio-economic structure; concepts used in economic decision making.

cularly necessary for the analysis of the mechanics of exhaustion of soil, deterioration of the vegetation, and the breakdown of systems of agriculture and stock rearing.

b) Research into possible alternatives in that environment and the spheres in which to act - geographical spheres certainly, but equally social, economic and political. In this research, one could test the validity of proposed techniques of analysing change in an eco-system which have already been tried on a smaller scale (Langley, 1975; Bugnicourt and Langley, forthcoming).

c) Formulation of a global ideology of social change and transformation of this environment: the definition of 'towards what' the population is aiming and its implications for the eco-system.

If one examines the social processes resulting from development projects, one can identify, in many cases, mechanisms of domination founded on the global social ideology of the 'experts' and decision makers, and using concepts and ideas completely foreign to the population which is the object of the 'development' in question. (It is interesting, in this respect, to compare the meaning in different cultures of the translations of words such as 'development', 'economic growth', 'needs', 'progress'...)

d) Identification and formulation of projects by the people, notably in matters concerning their spatial aspects and their effect on the different factors making up the environment. The use of the group's own concepts will certainly be a prelude to the flowering of inventiveness and creativity in the conception and execution of projects by the people.

e) 'Coordination' of planning, that is, of the steps necessary to assure the coherence of socio-spatial groups taking account of c) above. This is particularly important wherever changes are taking place in the use of rural space, as for instance in the case of agricultural reforms. One might wonder whether socio-spatial units should be identical throughout a national territory. There is room for doubt, simply because of the diversity of natural conditions, but one may suppose that historical criteria proper to each eco-culture have given rise to the growth of different kinds of units.

The use of space among the Fula pastoralists of Ferlo is far from identical to that of the rice-growing Diola of Lower Casamance, and the differences are reflected, according to all the evidence, in the spatial concepts and even the concepts of land use and socio-political organisation of their respective societies.

But there is no question here of proposing a return to 'traditional' categories and use of space and the environment. This would be a dead end and in any case, a false trail, as traditional society no longer exists.

> 'Where are the "traditional" societies of today? Fulfilling the function of providing cheap labour to the "modern" sector, the majority society, from the point of view of the number of men it includes, the so called "traditional" society, no longer is so: it is pseudo-traditional, it has a transformed, deformed, submissive traditionalism... Phenomena too often analysed by

sociologists as "survivals" - such as "tribalism" or village and family, clan and ethnic solidarities among urban immigrants - are ... pseudo-traditional forms hiding a "modern", if poor, reality; .. they are a means of survival in the dramatic conditions of "marginality"' (Amin, 1972).

Here we are advocating rather the full use of the knowledge of the environment of the great majority of the rural population in order to find, in cooperation with them, a form of land use which will be beneficial to them.

f) Training must be linked to planning and projects aimed at transforming the environment. Such action, where 'economic development and training would henceforth be woven into the same fabric' (Bugnicourt, 1973), by integrating training with 'a series of urgent technological moves in the fields of development and planning' and with the work itself, can only benefit from being explained in terms comprehensible to - and hence easily mastered by - the interested parties.

REFERENCES

ABASIEKONG, E.M. 1974 'Convincing farmers to use fertiliser alone is not enough', *Samaru Agricultural Newsletter*, 16, 49-51.

ALBENQUE, A. 1965 'Les cartes du terroir d'Etyolo: village bassari', *Bull. et Mém. de la Société d'Anthropologie de Paris*, 8, XIe série, 45-74.

AMIN, P. 1972 'Sous-développement et dépendance en Afrique Noire', *Partisans*, 64.

BARDIN, P. 1965 *La Vie d'un Douar*. Mouton, Paris and the Hague.

BARRAL, H. 1968 *Tiogo: Etude Géographique d'un Terroir Léla (Haute Volta)*. Mouton, Paris and the Hague.

BEARDSLEY, R.K., HALL, J.W. & WARD, R.E. 1959 *Village Japan*. Chicago.

BIEBUYCK, D. 1963 'Problemes d'analyse et de terminologie', in *African Agrarian Systems*. OUP *for* IAI, London. 1-19.

BOHANNAN, P. 1963 ' "Land", "tenure" and "land tenure",' in D. Biebuyck (ed.) *African Agrarian Systems*. 101-111.

BORDIEU, P. 1970 'La maison Kabyle ou le monde renversé', in *Echanges et Communications. Mélanges offerts à Claude Lévi-Strauss à l'occasion de son 60e anniversaire*. Mouton, Paris and the Hague.

BOUGHALI, M. 1974 *La Représentation de l'Espace chez le Marocain Illétré. Mythes et Tradition Orale*. Anthropos, Paris.

BOUTILLIER, J.L. et al. 1962 *La Moyenne Vallée du Sénégal*. PUF, Paris.

BOUZOUBAA, A. et al. 1974 *Occupation des Pièces dans une Habitation Traditionnelle au Maroc*. Ecole d'Architecture de Rennes. 47p. mimeo.

BUGNICOURT, J. 1973 'Pour l'Aménagement des Campagnes Africaines: une Formation Action', *Tiers-Monde*, 14, (54).

—— & LANGLEY, P. forthcoming. 'L'utilisation des schémas prospectifs pour l'analyse d'un écosystème.'

DALBY, D. 1964 'The noun *gàrii* in Hausa: a semantic study', *Journal of African Languages*, 3, 273-305.

DECORNAY, J. 1975 'Un nouveau Vietnam..' *Le Monde*, 22 April.

DE SCHLIPPE, P. 1956 *Shifting Cultivation in Africa*. London.

LANGLEY, P. 1975 *Analyse du Processus de Changement de l'Environnement Rurale en vue des Actions de Développement* (an account of field work presented to the Journées de Réflexion sur l'Environnement Rural et l'Espace Habité, Training for the Environment Programme, Dakar, February).

MAAMOURI, M. 1972 'Ethnolinguistique et alphabétisation', *Revue Tunisienne des Sciences Sociales*, 28-29.

PETTONET, C. 1972 'Espace, distance et dimension dans une société musulmane. A propos du bidonville de Douar Doum à Rabat', *L'Homme*, 12, 47-84.

REMY, G. 1967 *Yobri: Etude Géographique du Terroir d'un Village Gourmantché de Haute Volta.* Mouton, Paris and the Hague.

RICHARDS, P. 1975 'Drought in the Sahel', *African Environment/Environnement Africain*, I, 2. 3-6.

SACHS, I. 1971 *La Découverte du Tiers-Monde.* Flammarion, Paris.

SCUDDER, T. 1962 *The Ecology of the Gwembe Tonga.* Manchester UP and Rhodes-Livingstone Institute.

SIRVEN, P. 1970 'Les villages suburbains de Bouaké en voie d'urbanisation', in:*La Croissance Urbaine en Afrique Noire et Madagascar.* CNRS, Paris.

SNYDER, F.G. forthcoming. 'Banjal-Diola space categories: some aspects', *African Environment/Environnement Africain.*

TAIT, D. 1961 *The Konkomba of Northern Ghana.* OUP *for* IAI, London.

UNESCO. 1971 'Construction des Modèles d'Ecosystèmes', in:*Conseil International de Coordination du Programme sur l'Homme et la Biosphère. Première Session - Rapport Final.* MAB, UNESCO, Paris. 43-8.

'ALTERNATIVE' STRATEGIES FOR THE AFRICAN ENVIRONMENT

'Folk ecology', as a basis for community orientated agricultural development.*

PAUL RICHARDS
University College, London

Introduction

Our crudely materialistic society tends to view the natural world as an artefact-like entity set aside for man's sole use and enjoyment. The terms we use - habitat, resources, environment - are indicative both of the alienation we feel and the sense of ownership we wish to assert in regard to the world about us. This view of nature has penetrated to the core of present development theory. Thus we find that 'environment' is on the one hand a focus of intense interest because we see or assume a vital connection between agricultural and mineral resources and economic development, and on the other hand something of a scapegoat, because on the grounds of both political expediency and intellectual conceit we find it difficult to blame ourselves and the societies in which we live for the manifest failings of our economic systems. It is much more attractive to blame the continuing poverty of the Third World on, for example, 'the intrinsic infertility of tropical soils' or on 'the repeated occurrence of climatic catastrophes.'

There are three important corollaries of this aggressively positivistic (or perhaps we should say negativistic?) view of environment as a factor in development, namely, i) that there is a single 'best' or 'right' way to make use of a given natural environment, ii) that the search for this high road to prosperity requires substantially the same kind of education, science and technology employed in industrially advanced societies and, conversely, iii) that the persistence of traditional (and supposedly irrational) attitudes to, and practices of, environmental management can seriously inhibit progress. This view, an integral part of many theories of 'modernisation', can be tagged for convenience with the label 'scientism'. The foundations for an attack on scientism were laid long ago by Kant and Marx. Kant was concerned to justify the world view associated with Newtonian physics. He found he could not do this on logical grounds alone, and concluded that if Newtonian physics was necessary then the basic geography of the universe associated with that physics must also be necessary, *a priori* and intuitively. This notion that Newtonian physics is grounded in essential but unprovable assumptions helped prepare the way for the idea that all world views and their associated geometries are ultimately a function of the choice of a set of axioms. This choice of one set of axioms over another is to some degree arbitrary (though we may be victims of our own molecular biology in a number of respects, cf. Monod, 1972) and the competing claims of rival cosmologies are not to be

*English original

assessed in terms of which is 'correct' but which is best or most appropriate for a stated purpose. The Newtonian cosmos therefore is not 'disproven' by Einsteinian theory but simply takes its place as a scale-specific 'best approximation' in a more general formulation of the problem - appropriate in some contexts but not in others.

We are thus made aware of a degree of epistemological relativity in the way we formulate theories about the world which should guard us from making over-simple and invidious distinctions between magico-religious modes of thought on the one hand - which Lévy-Bruhl (1910) for example would suggest are pre-logical - and scientific modes on the other. The value of any given cosmology can only be assessed in relation to its purposes. Einsteinian theory is appropriate to sub-atomic particles and galaxies but is irrelevant to the solution of social questions. Zande witchcraft is a perfectly logical and effective - if symbolical and theoretical - way of coping with and explaining problems and pressures of the social cosmos. Not that logic and effectiveness are the only criteria for judging theory. In some respects the sociological theories of Hitler were perfectly logical and perfectly effective as well as being perfectly awful. The main point is, however, that judged from the standpoint of what they are designed to achieve non-western cosmologies are not innately inferior. They are systematically structured and operate with theoretical entities which are comparable in some respects to the abstractions of modern physics (cf. Horton, 1967). Furthermore the fact of their being couched in the language of religious symbolism rather than that of materialistic 'positivism' is not necessarily inappropriate to economic development, given that many non-western cosmologies explicitly seek to incorporate man as rational decision-taking moral being into their picture of the universe. Western thought on the other hand has striven to separate religion from science. How successful this has been is a moot point and we might wish to change the emphasis of Horton's conclusion that the gods and spirits of African religion are comparable to the unseen elementary particles of modern physics by suggesting that the physicist's belief in the existence of elementary particles has more in common with the worship of spirit forces than we would at first imagine. But whatever the case, the attempt to expunge religious elements from scientific dialogue has had the overall effect of reducing man-in-the-cosmos to the status of a material object devoid of any moral or ethical significance or responsibility. The end result in regard to social problems can be glimpsed in the disastrously dehumanised social engineering of B.F. Skinner's 'operant behaviourism' (cf. Skinner, 1969).

The argument so far has been a negative one. It is not seriously suggested that witchcraft beliefs and spirit cults constitute viable bases for contemporary cosmology or a valid frame of reference for development theory, even though in some respects, if welfare and survival are criteria of effectiveness, they may have points to commend them over the world-views of certain war-mongering technocrats. The intention rather is to suggest that an exaggerated respect for western science and technology and a crudely materialistic approach to the natural world may be dangerous foundations for effective development in Africa.

To take a more positive stance it is interesting to begin by considering some notions attributed to Karl Marx. Marx drew attention to the network of relationships and mutual influences which link our ideas about society and the world in which we live to technological progress and social change. The fact that cosmological ideas and theories may be influenced, in part at least, by social situations makes it difficult to assess their worth by conventional means. 'Environmental science' is partly concerned with understanding phenomena, partly with action, partly with things, partly with people. In this hybrid situation, facts - the stereotyped universals of positivistic science - can be dangerously misleading because the significance of empirical events and the relevance of ideas is determined by a context which is historical and therefore always changing. The forms of social organisation - from small hunting bands to modern industrial society - are rarely if ever perfectly balanced and at rest. Conflict, i.e. struggle for power, is an element in all societies, and as this struggle progresses new forms of social organisation (and new attendant philosophies) emerge. Marx suggests that the only society with long-term stability is one based on communal principles in which each individual, of his own free will, seeks his neighbour's welfare. If we are to survive at all this must be considered, by analogy with statistical usage, the most probable state of society. Given this notion of history as process it is not a question of 'facts' being 'true' or 'false' but of ideas and events being 'good' i.e. progressive or 'bad' i.e. reactionary, according to whether they reinforce or retard the historical trend (notwithstanding the likelihood that this trend may have more wrinkles, reversals and diversions than the nineteenth century may have supposed). 'By their fruits ye shall know them' would appear to be the appropriate standard to adopt.

It should perhaps be noted in passing that there is a degree of vagueness about the final vision. It is not clear how self-centred, self-seeking man is to be transformed. New moral and ethical standards are pre-supposed but do not appear to emerge from history as inevitably as, say, modern industrial societies have emerged from feudal contexts. Certainly one could argue that socialism in practice is often characterised by a lack of precisely those value which it is implied are vital to the emergence of a truly harmonious and egalitarian society. Che Guevara is reputed to have remarked that the revolution in Latin America would come only when the Marxists and Christians joined their forces. Marxist concern for social justice allied to the religious vision prominent in Christian and Muslim belief of God-in-History working to transform men's lives in moral and ethical terms would certainly present exciting possibilities for bringing into focus some of the key issues in the economic development of the Third World. These issues, however, lie outside the range of the present discussion. For the moment I am concerned with the following basic points:

i) The exclusion of religious elements from environmental discourse, far from being an advantage, may be interpreted as evidence of a declining concern for social as opposed to material welfare. To the extent that current so-called 'scientific' approaches to resource management stress profit rather than welfare, and exploitation rather than sharing, they are arguably inappropriate as bases for truly progressive development planning. Alternative attitudes may have to be developed therefore.

ii) Any set of ideas or attitudes to the natural world which runs counter to the prevailing greed and selfishness or which highlights the need for a more just and egalitarian society based on an active concern for the welfare of all must, conversely, be considered progressive and valuable and should therefore be encouraged. African traditional cosmologies often embody notions which far from being 'illogical and anti-scientific', and therefore 'barriers to progress', serve to protect the community and its lands from the depredations of the individual and so act in the long-term interests of environmental conservation. While it is true that these cosmologies are being eroded by the newer capitalistic religion of commodity fetishism it is important, and therefore a matter of urgency as well as importance, that their truly progressive elements should be identified and given due recognition, in order that they might make an effective contribution to the world-wide debate concerning the need for new attitudes towards, and alternative educational standards in, environmental resource management (cf. Schumacher, 1973).

'Folk Ecology' - Four Areas of Enquiry

Despite the rather overblown philosophical introduction, my purpose in this paper is entirely practical, for I am convinced that the culture of the ordinary people together with the experience they have gained coping with the day-to-day problems of contemporary economic and social change contain ideas and techniques which could be of immediate value in the struggle to improve agricultural yields and farm incomes. The full benefit of using the best of these ideas and techniques is only likely to be felt where a scaled-down grass-roots approach to development planning has been adopted, as for example in Tanzania. But even for countries pinning their major hopes to the 'Big Science-Big Business' equation it would still be a good idea to collect the basic information in order to keep the option open. Thus a country like Nigeria, while pursuing a vigorous programme of technological modernisation and rapid industrialisation, is nevertheless wisely devoting some resources to the study of traditional funds of scientific knowledge - for example the work being done on traditional pharmacopoeias at the University of Ife - as part of a more general strategy of cultural revival. This paper reviews research in the field of traditional agricultural knowledge and 'folk ecology' therefore, and also presents some preliminary results from various projects of this kind that I have been involved with, in order to try and stimulate a wider interest, perhaps, in this most important area of concern for the African environment.

One can identify, roughly speaking, four major areas of research interest in folk ecology. To begin with there are a number of studies by anthropologists and ethnographers of traditional cosmological and religious ideas (cf. Evans-Pritchard 1937; Diamond 1964a;1964b; Griaule 1965; Forde 1965). In most cases these adopt a 'functional' viewpoint and focus on those elements in traditional world views which contribute to the maintenance of group stability and coherence. But this deliberately contrived emphasis on continuity rather than change - and the fact that field work for many of the most famous case studies was done 20-30 years ago - means that much of the material has little practical relevance to agricultural development today. Even so it would be wrong to dismiss the entire field out of hand. There are papers by Horton (1962) and Douglas (1972), for example, which have something to say to even the most practically minded extension worker.

(A further problem would be that, in general, anthropologists writing about traditional cosmological ideas are much more concerned with people than with the soil and sky - that is, they pay more attention to the social than to the physical milieu. Again there are exceptions - eg. Bohannan's (1954) *Tiv Farm and Settlement* - but few other case studies of comparable environmental depth.)

When we turn to the closely related field of ethnolinguistics, however we run into a positive gold mine of useful material. Even straightforward lexical work on basic geographical or environmental terms can yield remarkable dividends - cf. Dalby's (1964) study of the Hausa term *gàrii*, generally glossed as town or sky, but in effect signifying the total human and physical environment of the Hausa city state. (In fact we badly need comparative studies of terms in African languages equivalent to English 'environment'/ French *'environnement'*.) But the full potential of linguistic investigation emerges only when we begin to consider complete terminologies. When for example we look not at isolated words for soil and plants but at whole sets of terms and complete taxonomies covering the biological domain. For it is when we study the principles governing the naming and classification of elements within the environment that we begin to understand how the traditional farmer conceptualises his world. These classificatory principles not only help us to see things as he sees them but also provide an insight into the fundamental process of linguistic and symbolic coding which permits a society either to replicate or to change its structure and culture from generation to generation (cf. Lévi-Strauss, 1962;1963).

The practical value of knowing how the farmer classifies his soil types, crops, weeds, medicinal plants and so on is readily apparent. A number of pioneering studies dealing with African terms for soils, climates and the seasons, have already appeared (Ojo, 1966; Netting, 1968; Benneh, 1969; Knight, 1974). The soil terminologies generally distinguish between different soil-types according to texture and/or colour - the latter generally being taken as surrogate for fertility (Aubert and Newsky, 1949). The distinctions are simple but effective from a practical point of view and in general equip the African farmer with adequate resources for an informed discussion of the land-use potential of his farm. The Ikale of southwestern Nigeria, farming sandy soils of uniformly low fertility, for example, distinguish only between hydromorphic and non-hydromorphic soils, but other groups of Yoruba farmers working in areas where fertility differentials are more marked and where soil catenas are strongly developed have more elaborate classifications (cf. Ojo, 1966). But *apropos* the Ikale it is perhaps worth noting that several of the farmers interviewed used the same term for fertiliser, pesticide and a witchcraft charm. This is where a sensitive extension worker might begin to anticipate trouble. The Ikale viewpoint is a perfectly understandable one given the background of their concepts of 'power' but it is equally apparent that recent attempts to introduce fertilisers in this area have not been helped by this element of terminological 'wire-crossing'. This kind of difficulty prompted Stone (1965;1967) to study Yoruba concepts concerning the physical world, including concepts of power, force, cause and effect. His study, which was aimed at helping the school science teacher, clearly has its weak points but it remains a fascinating pioneer effort. Several of the more problematic aspects of Stone's work concern the ways used in eliciting information, an area where there is a growing body of methodological literature (Colby, 1966; Hammer, 1966; Berlin, Breedlove and Raven, 1968).

There is also a growing debate on the related issue of whether, or to what extent, sets of concepts and terms in traditional thought can be considered 'scientific' (Horton, 1967; 1972; 1973; Emmet, 1972; Marwick, 1972; Pratt, 1972; Skorupski, 1973). With regard to classification Western science has long since abandoned the notion of a natural system. Classifications are created with specific purposes in mind and are meaningless, or at least misleading, unless this purpose is made explicit. Stone's work confirms that in a subsistence orientated society practical utility is an overriding consideration. Traditional classifications of plants and animals are no whit less logical than those of the academic biologist for stressing 'eatability' rather than philosophical principles derived from 'great chains of being' or theories of the evolution of species. It is possible, however, that the classificatory logic of village self-sufficiency may be at odds with the requirements of economic development. Whatever the case - whether the traditional terminology can be expanded to meet the needs of the new situation or whether the farmer must learn to handle a new set of terms and categories - it is evident that any deliberately induced programme of change must start with a thorough working knowledge of the traditional system of terminology (cf. Conklin, 1954). This calls for a certain amount of flexible and sympathetic collaboration between linguist, agronomist and ordinary farmer. It is also clear that this is a field in which African researchers will inevitably 'take the initiative'.

The third area of relevant research concerns decision making in peasant and subsistence economies. Several related lines of investigation have been pursued. First there is the theoretical work of economists and others on decision making criteria (e.g. Simon, 1957; and for a good general survey, Found, 1971). A number of studies have looked into the possibility that differentials in access to information - together with differences in the kind of information implicit in certain situations, cf. the distinction drawn between risk and uncertainty - are basic determinants of variations in environmental decision making under circumstances of rapid technological change. (cf. Gould, 1963; Hägerstrand, 1965a; 1965b; 1967; Chapman, 1974). Then there has been an interesting debate about the nature of economics in traditional societies. Colonial administrators developed a set of hypotheses about lazy farmers and target migrants unmoved by the normal considerations of material advantage and market economics (Hopkins, 1973). More recently aid-programme administrators have tended to explain away failures by pointing the finger of blame at 'irrational' cultural and religious barriers to the free flow of innovation. A certain amount of academic support for such ideas has been provided by those economists and anthropologists who would argue that the motivations of individuals as well as the whole character of economic transactions in pre-capitalist and non-monetised economies are radically different in kind from those prevailing in 'modern' economies (cf. Sahlins, 1960; Dalton, 1961). Others have strongly challenged this supposition and it would seem from evidence available (cf. Jones, 1960) that, whereas conventional textbook micro-economics is too narrow in scope to embrace the full wealth and variety of subsistence modes of production (cf. Clark and Haswell, 1964), irrational non-materially motivated *homo pre-economicus* is nothing more than a Piltdown sophistry, designed to create the impressing that capitalism and 'modern' economic development were brought to Africa by courtesy of European imperialists (cf. Crowder, 1968; Hill, 1963; 1970; Hopkins, 1973). Furthermore, several recent studies have shown that innovations are frequently rejected for the soundest of economic reasons - i.e. the new way is not as profitable as the old! (cf. Flinn, Jellema and Robinson, 1974) - and that where successful ideas do take hold they act in socially disruptive ways by increasing the income differential between the larger landowner and the 'middle peasant' who can afford new capital inputs, on the one hand, and the poorest farmers who cannot. It is this last circumstance which has given an ironic twist to the tale of success of the 'green revolution' in India (cf. King, 1973) and makes a more community orientated, 'tradition'-grounded, approach to agricultural development an alternative worth considering.

Finally there has been a certain amount - more would be welcome - of micro-level spade work by ethnobotanists, agricultural economists, geographers, etc. on African farming systems, showing that the basic agronomic practices of the African farmer are sound from the point of view of *long term* maximisation of output from the man-land ecosystem under prevailing technological and labour supply conditions, and assuming a stable population (cf. Blaut, 1961; Allan, 1965; Igbozurike, 1971). Even where population is rising it has been shown that there is much scope for 'indigenous adjustment' (cf. Mortimore, 1971; and studies in Prothero, 1972; and Zelinski, Kosinski and Prothero, 1970). Some have gone so far as to argue that a measure of population pressure is beneficial in inducing desirable technological developments in agriculture (cf. Boserup, 1965; Gleave and White, 1968). These kinds of argument are particularly important because previous arguments for an imported capital-and-technology-intensive agricultural revolution from outside Africa have always been based on the assumption of the utter inability of the traditional system to cope with rising population levels. This 'no choice' argument has been used to gloss over the blatant defects of 'imported' systems, namely:

a) monocultural dependence on an export crop;
b) dependence on imported inputs, eg. tractors, fertiliser, pesticide;
c) unevaluated ecological hazards associated with the above;
d) the social divisiveness of an 'advanced' system of production.

Once again the value of looking at alternatives is emphasised. The exact nature of the principles of environmental resource management employed by the African farmer are not yet fully understood. But work is being done on traditional practices of heaping, mulching, intercropping, multicropping, weeding and planting strategies, use of hydromorphic soils and the like which could constitute the basis for a truly satisfying alternative technology for agricultural development (cf. IITA, 1972;1973; Norman, 1974; Knight, 1974).

Folk Ecology and Agricultural Development

Technological innovation, mode of production, social organisation and beliefs are all interrelated. The search for an 'alternative' technology for agricultural development is long and difficult because it is not solely a technological matter. Questions about the kind of society in which such a technology would operate must be asked and answered. In the end the answers must come from the people themselves. It would be inappropriate, therefore, to pursue such issues here and now. What I would like to try and do however is to open up discussion on the role which the researcher might play in assisting this fundamental question and answer process. To do this I shall cite a number of examples - encountered in the course of fieldwork - of the sort of things which might have relevance, or are already proving of relevance, to community-based development efforts.

My own research aims, in a series of studies carried out in Nigeria over the last seven years, could be described as follows:

i) to document 'traditional' ideas about the agricultural environment;

ii) to develop a framework of analysis which could be used to distinguish a) those ideas which might be termed 'obstacles to change' from b) those which constitute realistic appraisals of specific environmental predicaments;

iii) to document the effort of those who believe in and actively try and develop (b)-type ideas for the community good - despite neglect and antipathy on the part of planners and researchers;

iv) to try as much as possible to see the farmer's environmental problems through his eyes.

There have been four projects. The first was a general study of tradition and change in Ikale farming. The second was a study of the farmer's response to an agricultural pest, the grasshopper *Zonocerus variegatus L.* (Jerath, 1965), carried out as part of the University of Ibadan/Centre of Overseas Pest Research, London, joint project on the biology and control of *Zonocerus*. This involved case studies from six different parts of Nigeria and also provided an opportunity to repeat some of the questions from the original Ikale survey three years later. The third project was a survey of farmers' attitudes and response to drought conditions in various parts of Nigeria done in collaboration with Dr. Julius Oguntoyinbo and the final project involved studies of peoples' perceptions of marketing, water-supply and farming problems in the Ibarapa area of Western Nigeria as part of a larger rural development planning project organised by the Geography Department, University of Ibadan (Daly, Filani and Richards, 1973).

In order to keep a little order in what can only be a most sketchy discussion I shall progress from an account of one or two interesting factual matters through mention of words, concepts, principles of folk taxonomy and the significance of complex symbolic forms such as proverbs, songs and religious ritual, to a discussion of the farmer's awareness of the processes and problems of his environment.

Factually speaking, farmers were often able to supply me with vast amounts of information about the utility of various plants (especially medicinal uses), the effects of different crop mixes, and the value of mulching, heaping, green-manuring with weeds, as well as making numerous shrewd observations about planting dates, impact of different climatic conditions on plant growth, soil fertility, texture, catena, hydromorphic properties and so on. (Ikale farmers were aware of why, in contrast to northern neighbours, they did not bother to categorise soils by colour contrast.) While much of this information and observation is common scientific knowledge, it is nevertheless quite possible that in some cases farmers have spotted hitherto undocumented environmental interrelationships, and have developed rule-of-thumb techniques for dealing with environmental problems which might provide interesting points of departure for research station investigation. The least that can be said is that an idea borrowed from the people, developed by the agronomist, and returned to the people again is much more likely to be adopted than something totally alien to the culture. A second point is that because farmers are close observers of weeds, pests, climate, etc. they can be a source of vital information which might otherwise be impossible to obtain. The Ikale questionnaire results, for example, yielded an impressive consensus about the years when *Z. variegatus* attacks first commenced (results checked out by the re-survey three years later) and allowed me to plot the year by year spread of the grasshopper from village to village in the area. Encouraged by this, we subsequently collected a good deal of oral historical data on the incidence of *Z. variegatus* attacks and data on droughts and famines for various parts of Nigeria (cf. Ajayi, 1962). Another instance where farmers were able to provide very useful orientational data concerned crops damaged by *Z. variegatus*. Laboratory information about feeding habits does not necessarily provide an infallible guide to economic damage caused by a pest of this kind. Much depends on the local land-use situation. Interviews with farmers gave us, first of all, some rough quantitative guidance as to the frequency with which various crops were attacked, and also some estimates of the relative impact on yield as opposed to straightforward vegetative damage. Melon in Ikale and coffee in Kabba both turned out to be major yield casualties though neither receives mention in reference handbooks such as Libby (1968). In addition, Ibarapa farmers, while agreeing that vegetative damage of cassava was often more spectacular than the ultimate yield loss, noted that the root from a grasshopper-defoliated plant is much harder than usual and cannot be used for making flour. These are all clues which could be followed up with profit under experimental conditions. Finally, and perhaps most surprising of all, it turned out that several Igbo groups ate *Z. variegatus* - sometimes in large quantities - and that far from wanting this grasshopper controlled as an agricultural pest some people wanted to know how to increase its numbers. Entomophagy is a badly neglected field of research in Africa (cf. Bodenheimer, 1951; Ene, 1963) but since it is possible that grasshoppers of various kinds are important supplementary elements in the diet of children there is surely a case for looking a little more closely at the nutritional virtues of this and similar 'casual' and 'emergency' food sources.

The material on terms, principles of taxonomy, proverbs and the like I am much more hesitant about and the pitfalls of interpretation are much more apparent to me than when I first started. Nevertheless a few generalisations and observations are worth risking for discussion purposes. It soon become apparent when compiling lists of names of plants and animals, for example, that many terms are of very restricted and local use. This was expecially true of insects, where names might change from village to village. Insect names were almost always descriptive (e.g. *Z. variegatus*, Alayi Igbo: *ogirise* = one that smells) and the richness of the terminology and indeed the ability to identify a given species with certainty correlated closely with its importance as a pest or its utility as an item of food. Standard sets of five different species of grasshopper were shown to farmers for identification and naming. The importance of *Z. variegatus* as an agricultural pest was underlined by the fact that in nearly all cases it was given a distinctive name, separating it from grasshoppers in general. But it was only in the various Igbo villages that local names were consistently given for each of the other four species. In nearly all cases each grasshopper was known for its food value, and informants could match the bottled specimens with freshly caught examples despite colour changes due to the preserving alcohol. In general the three Igbo dialects investigated (Alayi, Ibeku and Ngwa) appeared to have an almost inexhaustible richness of terms for what might be called the miscellanea of the agricultural environment - insects, weeds, edible minerals and clays, medicinal plants, etc. The Ikale by contrast - although over 90% of farmers named and identified *Z. variegatus* - responded to the other species Rorschach-test-fashion by confidently misidentifying *Gymnobothrus flaviventris* and *Eyprepocnemis Ibandana* as the yam cricket *Scapsipedus marginatus*[1]. This was, in fact, very useful as it drew attention to the severe damage done by this somewhat neglected pest (cf. Libby, 1968). As a result, farmers were asked to provide relative rankings of the importance of all other pests on their farms, and we can now place *Z. variegatus* in relation to these other pests in each area - information which could be useful in any future cost-benefit type review of entomological research priorities for Nigeria

Similar terminological work has been attempted for many other environmental parameters, including different types and periods of rainfall, drought, famine, planting season and so on. Again, one comes away with a feeling of the richness and detail of the farmer's view of his environment - and also a strong awareness that the distinctions he draws between different periods and seasons, while not always perhaps the same as those which the climatologists might favour, are nevertheless strongly rooted in the practical necessities of the agricultural system. Much the same can be said for the wisdom encapsulated in proverbs, chants, songs, and dramatic ritual. Proverbs are 'conservative' - frequently reinforcing the importance of traditional social values such as collaboration, as well as emphasising the agricultural virtues of timeliness and diligence. The Ikale say that 'he who plants water yam will regret in the end' - meaning that since the much more time-consuming white yam is so much better as food, laziness reaps its own reward. Similarly there is a proverb that points out that farming waits for no one, for 'when the *eki* tree *(Lophira alata)* has turned red it is already too late to start to clear a farm'. By contrast - because they are often composed for the occasion and move with the times - song texts are especially valuable as sources of information about changing attitudes and as an emotional commentary on how people perceive current problems (cf. Alagoa, 1968; Awe, 1974). In general, it could be said that students of environmental problems in Africa might find it an advantage to develop a much more sensitive literary and musical ear than hitherto.

[1]Special thanks to Bill Page for providing specimens and patiently identifying the numerous insects submitted by farmers in response to questions on the comparative importance of *Zonocerus*.

A good deal of attention was devoted to investigation of how farmers formulated, ranked and accounted for their problems. Apart from again providing the scientist with some guide as to what the villager is most concerned about - and it soon became clear that the villager's list of research priorities would be considerably at variance with those which might be gleaned from research station annual reports - the data helped enormously in the task of getting under the skin of rural problems. For a start the frequent use of religious and eschatological language helped emphasise the great extent of the desire for social and ethical reform. Secondly, farmers - as is to be expected in situations where subsistence elements are still important - tended to focus on 'ecological' rather than 'economic' problems - on what to do about pests, weeds, soil erosion etc. rather than on transport and marketing difficulties, unfair prices, etc. The situation varied from place to place, of course, and in areas with a major cash crop or closer to big cities people were much more preoccupied with economic aspects of the situation. There was also much greater emphasis on economic difficulties when comparing the responses of Ikale farmers in 1974 with those they had given in 1971. But all in all this suggests that one possible 'barrier to change' might be a degree of misidentification whereby basic difficulties are diagnosed in much too narrow biological terms rather than in a broader economic perspective. This in turn suggests that a major preoccupation of the extension service should be the dissemination of much more 'economic' information - regular market price reports over the radio would be a start - while indicating at the same time that farmers will be especially amenable to the adoption of 'ecological' innovations - fertilisers, new seeds, etc. The preoccupation with pests and weeds would seem to indicate that pesticides and herbicides are likely to be the most spontaneously attractive innovations in the near future as far as farmers in Nigeria are concerned. There is independent evidence in the Ikale study that pesticides are much more in demand than fertilisers.

This leads us finally to ask what kind of understanding farmers have of environmental processes - especially some of the more problematical processes which have accompanied economic development. The answer would seem to be a very good understanding of those processes which lie within the scope of the individual to observe from start to finish. Here farmers often draw extremely shrewd inferences. Alayi farmers, for example, cite thunder as the cause of the cassava blight epidemic. In fact it does appear to be spread by heavy rainfall. Ikale farmers have a traditional explanation for rainfall which highlights the causal role of bush burning. Climatologists confirm that smoke in the atmosphere, by providing numerous condensation nuclei, increases the chance of rainfall. A fair number of Ikale farmers were also able to point out the relationship between over-cultivation and increases in the number of weeds and pests, and one acute observer noted that roads assisted the spread of weeds.

The *Z. variegatus* survey provided a good opportunity to find out just how well farmers could analyse what was in essence a new and potentially catastrophic situation and how they could then put this knowledge into practice. For a start there was much evidence of accurate and painstaking observation. Most people remembered invasion dates and periods, and had a good idea of the life-cycle of the insect. Some even noted when eggs were being laid and the periods when the grasshopper was parasitised by *Blaesoxipha filipjevi* (this was the close-season for eating the grasshopper as far as people in Alayi were concerned, though the Ngwa apparently eat them whatever, parasites and all). Sensible inferences were also widely drawn. Innumerable respondents associated the explosive rise in *Z. variegatus* populations with the recent rapid spread of the weed *Eupatorium odoratum* (Akintola weed in Ikale, and Awolowo weed in Kabba according to political preferences!). It is in fact established that *E. odoratum* is good feeding for young grasshoppers and the farmer's correlation may, therefore, have a deal of validity (COPR, 1974). Similarly, several farmers suggested a correlation between the severity of the dry season and the population fluctuations which are one of the most puzzling aspects of the grasshopper's ecology. Once again the signs indicate that this is an entirely valid inference (COPR, 1974). Alayi children know that the time to catch *Z. variegatus* is early in the morning because it is more sluggish then than later in the day, and finally one inspired soul gave a basic reason for the severity of the attack on cassava when according to laboratory results this is by no means a favourite food - namely that cassava is one of the very few plants with green leaves surviving in farm fields during the dry season (cf. COPR, 1974; 1975).

The adaptations and adjustments that people were prepared to make in the light of inferences such as these were equally sensible. Alayi farmers made extensive use of ashes as an insect repellant (presumably a cassava leaf liberally garnished with wood ash is rendered less palatable) though they were prepared to admit that this was not as effective as they would have liked it to be. Quite a number of farmers had experimented with Gammalin spraying (especially in cocoa growing areas where this treatment is familiar already). Gammalin is, in fact, the insecticide indicated as a result of COPR/UI trials (COPR, 1975). Some Igbo farmers set out to reduce *Z. variegatus* populations with locally made 'insect traps'. Others in despair had changed their crop patterns and cut down on the crops worst affected, while several Kabba farmers have backed up their belief in the causal role of *E. odoratum* by cutting this weed out of their farms where-ever they could.

But if we leave biological examples such as grasshopper problems and turn to consider economic processes the farmers' understanding was often less good, largely because so much of economics is determined and decided in places the villager can hardly ever hope to visit. Our Ibarapa study paid particular attention to the way in which people explained economic processes (cf. Filani and Richards, 1975). Price fluctuations were accurately accounted for when commodities having a local market were examined. There was a solid appreciation of the way poor communications forced up the price to the consumer, and also of the way lack of assemblers and bulkers could result in sellers competing amongst themselves to off-load farm produce and so bidding down the price. Supply fluctuations were given as the major reason for seasonal price variations, but demand factors were cited much less frequently, largely because the effective demand for Ibarapa farm produce is far away in Lagos and Ibadan. Farmers admitted that in many cases they had no means of knowing the Ibadan price for their goods and suspected that they were being cheated by traders. The greatest difficulties were experienced - not surprisingly - when it came to accounting for the complex behaviour of cocoa prices. Most farmers seemed to assume that cocoa marketing simply reflected governmental perfidy. In other words farmers understand market mechanisms and will react perfectly rationally to economic situations in which they have some measure of say, but feel at a loss to account for those apparently arbitrary processes which have their focus far beyond the limits of the villager's mental map.

Conclusion

There are two straightforward points I want to emphasise by way of conclusion. The first is that there is much scope for improved communication and collaboration between biological and environmental scientists on the one hand and linguists and social scientists on the other as far as research in Africa is concerned. The second and more fundamental point is to reiterate that the peasant farmer is an acute observer and shrewd analyst of his own difficulties. Community mobilisation for progressive agricultural change based on 'folk ecology' is, therefore, not necessarily a sentimentalist's pipe dream. But even if it proves impossible to bring the university to the village, Chinese style, there should nevertheless be much more of a two-way flow of information between agricultural researchers on the one hand and ordinary farmers on the other if the problems of the African agricultural environment are to be dealt with in an effective manner. Big Science has had its chance - and for every ERTS satellite has given us the dubious blessings of several MIRVS and MARVS. Perhaps it is now time to stand up and be counted for shoe-string budgeted, action-orientated, peace-focused and food-first people's science.

Acknowledgements

I would like to acknowledge a Goldsmith's Travelling Scholarship, a grant from the University of Ibadan Senate Research Fund and funding from the COPR/UI *Z. Variegatus* Study as support for the various research projects referred to in this paper. I would also like to say thank you to all my field assistants and to the very many farmers who have responded to our questions and shown such courteous and encouraging interest in the well-being of the various projects. Special thanks to Johnson Ogunji for teaching me more about farming, and life in general, than I ever had a right to expect.

REFERENCES

AJAYI, J.F.A. 1962 *Population Census of Nigeria. List of Historical Events for Determination of Individual Ages.* Ibadan.

ALAGOA, E.J. 1968 'Songs as historical data. Examples from the Niger Delta', *Research Review,* Inst. of African Studies, University of Ghana, 5, 1-16.

ALLAN, W. 1965 *The African Husbandman.* London.

AUBERT, G. & NEWSKY, B. 1949 'Note on the vernacular names of the soils of the Sudan and Senegal',Technical communication of the Commonwealth Bureau of Soil Science 46, Harpenden. 107-9.

AWE, B. 1974 'Praise poems as historical data: the example of the Yoruba *oriki*', *Africa,* 44, 331-49.

BENNEH, G. 1969 'The attitudes of the Kusasi farmer of the upper region of Ghana to his physical environment', *Research Review,* University of Ghana, Institute of African Studies, 6, 87-100.

BERLIN, B., BREEDLOVE, D.E. & RAVEN, P.H. 1968 'Covert categories and folk taxonomies', *American Anthropologist,* 70, 290-99.

BLAUT, J.M. 1961 'The ecology of tropical farming systems', *Revista Geografica* 28, 47-67.

BODENHEIMER, F.S. 1951 *Insects as Human Food.* The Hague.

BOHANNAN, P. 1954 *Tiv Farm and Settlement.* London.

BOSERUP, E. 1965 *The Conditions of Agricultural Growth: the Economics of Agrarian Change under Population Pressure.* London.

CHAPMAN, G.P. 1974 'Perception and regulation: a case study of farmers in Bihar', *Trans. Inst. Br. Geogr.,* 62, 71-94.

CLARK, C. & HASWELL, M. 1964 *The Economics of Subsistence Agriculture.* London.

COLBY, B.N. 1966 'Ethnographic semantics: a preliminary survey', *Current Anthropology,* 7, 3-32.

CONKLIN, H.C. 1954 'An ethno-ecological approach to shifting agriculture', *Transactions*, New York Academy of Sciences, 2nd. Series, 17, 133-142.

COPR 1974 *Control of Zonocerus Variegatus L. in Nigeria: Interim Report January 1972 - July 1973.* Centre for Overseas Pest Research, London.

—— 1975 *Control of Zonocerus Variegatus L. in Nigeria: 2nd Interim Report, 1973-74.* Centre for Overseas Pest Research, London.

CROWDER, M. 1968 *West Africa under Colonial Rule.* London.

DALBY, D. 1964 'The noun *gàrìi* in Hausa', *Journal of African Languages,* 3, 273-305.

DALTON, G. 1961 'Economic theory and primitive society', *American Anthropologist,* 63, 1-25.

DALY, M.T., FILANI, M.O. & RICHARDS, P. 1973 'The Ibarapa regional planning profile', Geography Dept., University of Ibadan. mimeo.

DIAMOND, S. 1964a 'Anaguta cosmography: the linguistic and behavioral implications', *Anthropological Linguistics,* 2, 31-38.

—— (ed.) 1964b *Primitive Views of the World.* New York.

DOUGLAS, M. 1972 'Environments at risk', in: J. Benthall (ed.) *Ecology, the Shaping Enquiry.* London, 129-145.

EMMET, D. 1972 'Haunted universes', *Second Order,* 1, 34-42.

ENE, J.C. 1963 *Insects and Man in West Africa.* Ibadan.

EVANS-PRITCHARD, E.E. 1937 *Witchcraft, Oracles and Magic among the Azande.* Oxford.

FILANI, M.O. & RICHARDS, P. 1975 'Roads, markets and rural development: a case study from western Nigeria'. Geography Dept., University of Ibadan. mimeo.

FLINN, J.C., JELLEMA, B.M. & ROBINSON, K.L. 1974 'Barriers to increasing food production in the lowland humid tropics of Africa'. Paper read at Seminar of OLAC/MEA Program Advisors in Agriculture, Ibadan, 1 May 1974.

FORDE, D. (ed.) 1965 *African Worlds: Studies in the Cosmological Ideas and Social Values of African Peoples*. London.

FOUND, W.C. 1971 *A Theoretical Approach to Rural Land Use Patterns*. London.

GLEAVE, M.B. & WHITE, H.P. 1968 'Population density and agricultural systems in West Africa', in: M.E. Thomas & G.W. Whittington, *Environment and Land Use in Africa*. London.

GOULD, P. 1963 'Man against his environment: a game theoretic frame-work', *Annals of the Association of American Geographers*, 53,290-97.

GRIAULE, M. 1965 *Conversations with Ogotemmêli: an introduction to Dogan religious ideas*. London.

HÄGERSTRAND, T. 1965a 'Aspects of the spatial structure of social communication and the diffusion of information', *Papers of the Regional Science Association*, 16, 28-42.

—— 1965b 'A Monte Carlo approach to diffusion', *Archives Européennes de Sociologie*, 6, 43-67.

—— 1967 *Innovation Diffusion as Spatial Process*. (trans. A.R. Pred). Chicago.

HAMMER, M. 1966 'Some comments on formal analysis of grammatical semantic systems', *American Anthropologist*, 68, 362-73.

HILL, P. 1963 *Migrant Cocoa Farmers of Southern Ghana*. Cambridge.

—— 1970 *Studies in Rural Capitalism in West Africa*. Cambridge.

HOPKINS, A.G. 1973 *An Economic History of West Africa*. London.

HORTON, R. 1962 'The Kalabari world view: an outline and interpretation', *Africa*, 32, 197-220.

—— 1967 'African traditional thought and Western science', *Africa*, 37, 50-71; 155-87.

—— 1972 'Spiritual beings and elementary particles - a reply to Mr. Pratt', *Second Order*, 1, 21-33.

—— 1973 'Lévy-Bruhl among the scientists - a reply to Mr. Skorupski', *Second Order*, 2, 14-30.

IGBOZURIKE, M.U. 1971 'Ecological balance in tropical agriculture', *Geographical Review*, 61, 518-29.

IITA, 1972 *Report: Farming Systems Program*. International Institute of Tropical Agriculture, Ibadan.

—— 1973 *Report: Farming Systems Program*. Ibadan.

JERATH, M.L. 1965 'Notes on the biology of *Zonocerus variegatus Linnaeus* from Eastern Nigeria', *Revue Zool. Bot. Afr.*, 72, 243-51.

JONES, W.O. 1960 'Economic man in Africa', *Food Research Institute Studies*, 1, 107-34.

KING, R. 1973 'Geographical Perspectives on the Green Revolution: a Review Article'. *Tijdschrift voor Economische en Sociale Geografie* 64, 237-44.

KNIGHT, C.G. 1974 *Ecology and Change: Rural Modernization in an African Community*. New York.

LEVI-STRAUSS, C. 1962 *La Pensée Sauvage*. Paris.

—— 1963 *Structural Anthropology*. New York.

LEVY-BRUHL, L. 1910 *Les Fonctions Mentales dans les Sociétés Inférieures*. Paris.

LIBBY, J.L. 1968 'Insect pests of Nigerian crops', *Research Bulletin*, College of Agricultural and Life Sciences,University of Wisconsin, 269.

MARWICK, M.G. 1972 'Anthropologists' declining productivity in the sociology of witchcraft', *American Anthropologist*, 74, 378-85.
MONOD, J. 1972 *Chance and Necessity*. (trans. A. Wainhouse). London.
MORTIMORE, M.J. 1971 'Population densities and systems of agricultural land-use in northern Nigeria', *Nigerian Geographical J.*, 14, 3-16.
NETTING, R. McC. 1968 *Hill Farmers of Nigeria*. Seattle.
NORMAN, D.W. 1974 'Crop mixtures under indigenous conditions in the northern part of Nigeria', *Samaru Research Bulletin*, 205, Institute for Agricultural Research, Samaru, Ahmadu Bello University, Zaria, Nigeria.
OJO, G.J.A. 1966 *Yoruba Culture*. London.
PRATT, V. 1972 'Science and traditional African religion', *Second Order*, 1, 7-19.
PROTHERO, R.M. (ed.) 1972. *People and Land in Africa South of the Sahara*. London.
SAHLINS, M.D. 1960 'Political power and the economy in primitive society', in: G. Dole and R. Carneiro (eds.) *Essays in the Science of Culture*. New York.
SCHUMACHER, E.F. 1973 *Small is Beautiful: a Study of Economics as if People Mattered*. London.
SIMON, H.A. 1957 *Models of Man*. New York.
SKINNER, B.F. 1969 *Contingencies of Reinforcement: a Theoretical Analysis*. New York.
SKORUPSKI, J. 1973 'Lévy-Bruhl among the scientists', *Second Order*, 2, 3-13.
STONE, R.H. 1965 *Yoruba Lore and the Universe*, Institute of Education, University of Ibadan, Occasional Publication no. 4, Ibadan.
—— 1967 *Yoruba Concepts of the Natural World in Relation to Learning Science*, Ph. D. thesis, London.
ZELINSKI, W., KOSINSKI, L. & PROTHERO, R.M. 1970 *Geography and a Crowding World*. New York.

INFORMATION FOR CONTRIBUTORS

The IAI will be pleased to hear from readers who wish to submit articles, to propose the publication of a relevant monograph or to suggest a topic for a Special Report. Any articles submitted will be considered in terms of their suitability for a projected IAI Special Report or the bulletin *African Environment*, published in Dakar. Articles concerning African societies and cultures may also be considered for the IAI's quarterly *Africa* and articles concerning language for the IAI's new annual *African Languages*. Contributors should bear in mind the wide audience, drawn from all academic disciplines, and from practical and administrative workers, to whom the Reports and Bulletin aim to appeal.
References should be given thus in the text: (Mambo, 1973); and thus in a list at the end of the article: MAMBO, P. 1973 'The effects of bush burning', *Africa*, 34, 106-33 *or* MAMBO, P. 1963 *Farming Techniques*. Nairobi.
Footnotes should be kept to a minimum.
Diagrams and maps should be clearly drawn on A4 size paper or larger and sources duly acknowledged.
Each contribution should be accompanied by an *abstract* of 150-300 words in length and by *details about the author* for a section of 'Notes on Contributors'.
In the interests of speed of production and because of the methods of printing at present used, it is not possible to provide *proofs* for authors.

ENVIRONMENTAL REVIEW UNIT
INTERNATIONAL AFRICAN INSTITUTE

JUNE
1975

PASTORAL NOMADS AND DEVELOPMENT

LES PASTEURS NOMADES FACE AU DEVELOPPEMENT

CLARE OXBY

A selected annotated bibliography with special reference to the Sahel, with an analytical introduction in English and French. The bibliography was compiled specifically to meet the practical needs of workers in this field of developme In her introduction the author discusses and evaluates the possible alternativ strategies for development both in the direction of the settlement of nomads a the modernisation of pastoralism.

35pp *Price : 80p.*

Already published

ABAAR

The Somali Drought

Edited by I.M.Lewis

The first Emergency Report of the Environmental Review Unit combining up to date statistical assessment of the drought with informative background material and vivid eye-witness accounts.

44pp *4 maps* *80p.*

INTERNATIONAL AFRICAN INSTITUTE
210 HIGH HOLBORN LONDON WC1V 7BW

No. 1 (1975) in Press

A NEW ANNUAL JOURNAL

AFRICAN LANGUAGES

LANGUES AFRICAINES

Editors:
Pierre Alexandre
Ayọ Bamgbọse
Hazel Carter
K. Mateene
I. Richardson
A.K. Turay

Approx. 300pp. £5 per annum (indiv members £2.50, corporate members £

INTERNATIONAL AFRICAN INSTI
210 HIGH HOLBORN LONDON WC1